AGRICULTURAL INSECT PESTS AND THEIR CONTROL

Second Revised and Enlarged Edition

V.B. Awasthi
M.Sc., Ph.D.
Former Associate Professor
Department of Zoology
J.N.V. University, Jodhpur 342 001, India

SCIENTIFIC® PUBLISHERS
New Delhi • Jodhpur

SCIENTIFIC®
PUBLISHERS

Published by:
Jodhpur
5 A, New Pali Road, P.O. Box 91
Jodhpur - 342 001 Rajasthan (India)
E-mail: info@scientificpub.com
Website: www.scientificpubonline.com

New Delhi
4806/24, Ansari Road, Daryaganj,
New Delhi - 110 002 (India)

Print : 2024

First Edition: 2007
Reprinted : 2011
Reprinted : 2014
2nd Revised and Enlarged Edition : 2017
Reprinted : 2021
Reprinted : 2023

ISBN: 978-93-86237-12-5 (H/B)

e-ISBN: 978-93-87307-29-2

Printed in India

PREFACE
(2nd Edition)

I would like to thank both the teachers and the students of entomology for accepting whole heartedly the first edition of Agricultural Insect Pests and Their Control. In this, second edition, three new chapters viz., (1) Ecological tactics for management of insect pests, (2) Molecular techniques and habitat manipulation for parasitoid conservation in cropping systems and (3) Pests of Spices, Narcotics and Drugs have been added. Also at appropriate places in the text some minor corrections/alterations have been made. A large number of new illustrations have been reproduced along with original sources which were not available at the time of first edition. The selected bibliography has been updated. I hope the readers of agricultural entomology will find this edition of book more useful.

Jodhpur **Dr. V.B. AWASTHI**

PREFACE
(1st Edition)

This book has been written keeping in view the requirements of the students of Agricultural Entomology of various Indian Universities. The book provides a brief account on the structure, function, development and metamorphosis and a comprehensive account on almost all the methods of pest control (viz.: cultural control, physical and mechanical control, biological control including genetic sterile male methods, insect hormones as pesticides; plant quarantine and legal control, chemical control and integrated pest management). The various appliances used in the pest management have also been given. The nature of damage caused by the major insect pests of crops, their life histories and control that have been described are: Polyphagous pests, pests of cotton and fibre crops, pests of sugarcane, pests of oilseed crops, pests of pulse crops, pests of sorghum, pests of cereal crops, pests of fruits and fruit trees, pests of vegetables, pests of plantation crops, pests of soybean, insect and some other pests of floricultural crops and pests of stored grains.

Further, so far published books on Applied or Agricultural Entomology provide no information on the insect pests of soybean which has been considered as a "wonder crop" in India. Because it incorporates the qualities of pulses, oilseeds and vegetables and thus meeting important basic requirements of oil, protein and vegetables needed by man.

Some minor pests have also been described because at times, when the climatic conditions are favourable, their food habit changes and they may become injurious to several crops. It is interesting to note that the population dynamics and the outbreak of the insects in the pest form are directly influenced by (1) temperature, (2) humidity, (3) food, and (4) shelter in addition to several other factors which also control the survival and dispersal of pest species, in a particular area. This is very common in oligophagous and polyphagous insects.

Further, I would like to mention here that several chlorinated hydro-carbon insecticides like DDT, BHC aldrin, dieldrin etc, used in the past have recently been banned. In the control of pests, only those insecticides have been given that described in authentic books.

In the book, only from historical point of view, account of some of the chlorinated hydrocarbon-insecticides has been given under the footnote "banned". Likewise, several organophosphorus compounds, systemic insecticides and fumigants are also not in use in pest control now a days.

It is hoped that the students of Entomology of Agricultural Universities and Colleges will find the book extremely useful. The book will also be useful to the Plant Protection Departments and similar Institutions. A large number of figures have also been given to facilitate easy identification of the insect pests. In the appendixes the common names and trade names of some insecticides, as well as characteristics of insect orders and their agricultural importance have been given. Every attempt has been made to provide all the necessary information that is needed by the students of Agricultural Entomology and here I would like to cite Robert E. Pfadt (1962):

"As one may surmise from what has been said, text books are not the best places for anyone to seek the latest in insect control recommendations. On the other hand, texts do serve an important role in providing both a background of information on insects, their habits, life histories, and ecology and an appreciation of the insect problems on various crops and how we go about solving these problems."

In the course of compilation of the text material several books, journals, technical bulletins, research papers and review articles have been consulted. I express my sincere thanks to the authors, publi-shers, reviewers and scientists whose work form the base of the book.

I am thankful to my friend Dr. Allah Noor Khan, Associate Professor of Agricultural Entomology, A.R.S. Mandore, Jodhpur for going through the manuscript and offering valuable suggestions.

I am grateful to the authorities of Scientific Publishers (India), Jodhpur, Shri Prahlad Rai Sharma and Shri Pawan Kumar Sharma for their keen interest in my work and publishing the book nicely.

I am also extremely thankful to my wife Sneh Lata and son Dr. Santosh Kumar for their constant encouragement and whole hearted support.

I am particularly grateful to my daughter Shubha, for reading the proof and suggesting necessary corrections.

I would appreciate if readers of the book will inform me of any errors or omissions that may come to their notice.

Jodhpur **Dr. V.B. AWASTHI**

27th March, 2007

CONTENTS

1

INTRODUCTION

One cannot avoid seeing insects, as they are found everywhere, even in the Antarctic. They have been on this earth some 250–500 million years ago, and seem here to stay. More insects and more kinds of insects are known than all other animals visible to the naked eye. Insects have been called man's worst enemy. Certainly, they are, but some have economic value, and for other reasons we would be hard put to exist without them. Insects are gems of natural beauty, zoological mysteries, and a constant source of interest.

THE INSECTS

Insects are related to crabs and lobsters. Like these sea animals they possess a kind of skeleton on the outside of their bodies. The body itself is composed of three divisions: head, thorax, and abdomen. The thorax has three segments, each with a pair of jointed legs; so an insect normally has six legs. Most insects also have two pairs of wings attached to the thorax, but some have only one pair, and a few have none at all. Insects usually have two sets of jaws, two kinds of eyes– simple and compound – and one pair of antennae.

So much for the typical insects, but many common ones are not typical. The thorax and abdomen may appear to run together. Immature stages (larvae) of many insects are worm-like, though their six true legs and perhaps some extra false ones may be counted. Immature insects are often difficult to identify. It is also hard to tell the sex of some insects but in most cases, sexual dimorphism is distinct. In some groups males are larger or have larger antennae or different markings. The female is sometimes marked by a spear-like ovipositor for laying eggs extending from the base of the abdomen.

INSECT RELATIVES

A number of insect-like animals are confused with insects. Spiders have only two body divisions and four pairs of legs. They have no antennae. Some spiders do not live in webs. Other insect-like animals have the head and thorax joined like the spiders. Crustaceans have at least five pairs of legs and two pairs of antennae. Most live in water (crab, lobster, shrimp), but the lowly sowbug is a land crustacean. Centipedes and millipedes have many segments to their bodies with one pair of legs (centipedes) or two pairs (millipedes) on each. Centipedes have a pair of long antennae; millipedes have a short pair. Millipedes often coil up when distrubed.

NUMBER OF INSECTS

The insect group is by far the largest group of animals in the world. It has recently been estimated that the total number of insect species may range from 4-10 million (See: Atwal and Dhaliwal, 2012). The class of insects is divided into some 28 orders (depending on the classification). Five "major" orders stand out for their high species richness, the beetles (Coleoptera), flies (Diptera), wasps, ants, and bees (Hymenoptera), butterflies and moths (Lepidoptera), and the true bugs (Hemiptera). Beetles- appears to be confirmed since they comprise almost 40% of described insects (more than 350,000 species). The Hymenoptera have nearly 250,000 described species, with the Diptera and Lepidoptera having between 125,000 and 150,000 species, and Hemiptera approaching 95,000. Of the remaining orders of living insects, none exceed the 20,000 described species of the Orthoptera (grasshoppers, locusts, crickets and katydids). Dermaptera less than 2,000 described species and cockroaches (Blattaria) with only 4,000 species. Most of the "minor" orders have from some hundreds to a few thousands of described species (Gullan and Cranston, 2005).

SHAPE AND SIZE OF INSECTS

The shapes of insects vary very much. Some insects are spherical, others are flattened like a leaf or elongated like a stick and still others have bizarre shapes and are clothed with hairs, scales, spines, horns etc. Some insects are smaller than the larger Protozoa but others are larger than small mammals. Among the smallest insects are beetles, Trichopterygidae (0.5 mm long) and *Mymar* (0.2 mm long) parasitic in eggs of insects. Some beetles like Dynastidae, Cerambycidae etc. exceed the size of small mammals. The giant silk moth, *Attacus atlas* is found from India eastwards through China to Malaysia. It is reputed to be the largest moth in the world. It attains a wing expanse of 12 inches (Goodden, 1977, Anonymous, 2008). The Queen Alexandra's bird wing of Papua New Guinea is the largest butterfly which can be 12 inches across and the world's longest insect is the giant stick insect (Phasmida) of Indonesia which grows upto 13 inches long. (Anonymous, 2008).

INSECTS IN THEIR PLACE

In the broad view, insects play an important natural role, not only in ways that benefit man but as food for many kinds of fish, amphibians, birds, and mammals. Many of our songbirds depend almost entirely on an insect diet. Every fisherman knows how fresh-water game fish go after insects. Insects help make our rich plant life and wildlife possible. Keep this broad view in mind when people start talking about widespread insect control-something that may become possible with newer chemicals. Local control may be successful and useful to man, but control on a large scale might cause more harm than it would prevent, because insects are so important to most other kinds of life about us.

CONTROL OF INSECTS

There are ways to supplement the natural control of insects by birds and other animals. We encourage those harmless insects which prey on harmful kinds. We can exclude insects with screens, discourage them with repellents, trap them or poison them. Many of the newer poisons are highly effective. But since there are so many kinds of insects which live and feed in so many different ways, there is no single best method to get rid of them. Yet, with concerted effort some dangerous insects have been wiped out over fairly large areas. A unique example of this was the complete destruction of the Mediterranean fruit fly, *Ceratitis capitata.*

INSECTS AND MAN

Whether certain insects are to be considered helpful to man or otherwise, depends as much on man as it does on insects. Our ways of farming and raising animals have provided some insects which might otherwise be rare with conditions enabling them to multiply a thousandfold. Only about 1 per cent of insects are harmful, but these destroy about 10 per cent of our crops, causing a loss of approximately 4500-5000 million rupees annually. Other insects are parasitic on other animals, and some are known to carry diseases.

On the other hand, this would be a sorry world without insects. We would have no apples, grapes, or pear, much less cotton, and fewer oranges and garden vegetables for these and many other common plants depend on insects to pollinate their flowers. And there would be no honey, of course. Some insects aid the process of decay, a process that is essential to life. Some insects help to control others, and all help to maintain a balance in nature (see: Awasthi 2013).

2

INSECTS : STRUCTURE AND FUNCTION

Insects invade and thrive in more environments than any other group of animals. Although they essentially dwell on the earth's surface, they have also mastered the subterranean world and the realms of air and water. They inhabit deserts and rain forests, hot springs and snow fields, dark caves and sunlit surroundings. They eat the choicest foods on man's table and can eat the table too! It seems obvious that organisms with abilities such as these must have extreme structural and functional diversity. It is truly amazing that such diversity is found in animals as small as insects. The extreme range of environments utilized by insects is accompanied by an equal range of structural and functional variations (for details of this chapter see: Awasthi, 2005, 2009).

THE INTEGUMENT

The integument of the insects serves as exoskeleton and is ectodermal in origin. It performs several functions e.g. protection against moisture, dryness, diseases and natural enemies of many kinds. The integument also protects the soft tissues within, and serves, as the place of attachment for the numerous muscles. The integument consists of (1) cuticle, (2) epidermis or hypodermis and (3) basement membrane (Fig. 1–A).

(1) Cuticle : It is secreted by the underlying epidermis and is a non-cellular layer. It covers the whole of the body and its appendages. The cuticle also forms the lining of the fore-and hindgut, tracheal system and parts of the reproductive tract. The rectal glands and anal glands are also lined by the cuticle. The cuticle is divided into. (1) epicuticle and (2) procuticle. The procuticle is divided into exocuticle and endocuticle. The innermost layer is known as epidermis or hypodermis, which forms the innermost layer, lining of the body wall. It is the epidermis which is a cellular layer composed of various types of cells and their canals viz. tormogen cells, trichogen cells, dermal glands, oenocytes, pore canals, opening of dermal glands etc. The epidermis rests on a non-cellular basement membrane. In addition, a variety of cellular out growths viz. bristles, scales, sensory and glandular setae, clothing hairs and multi cellular out growths are present on the cuticular surface hence, it is never smooth. (Fig. 1 B,C,D).

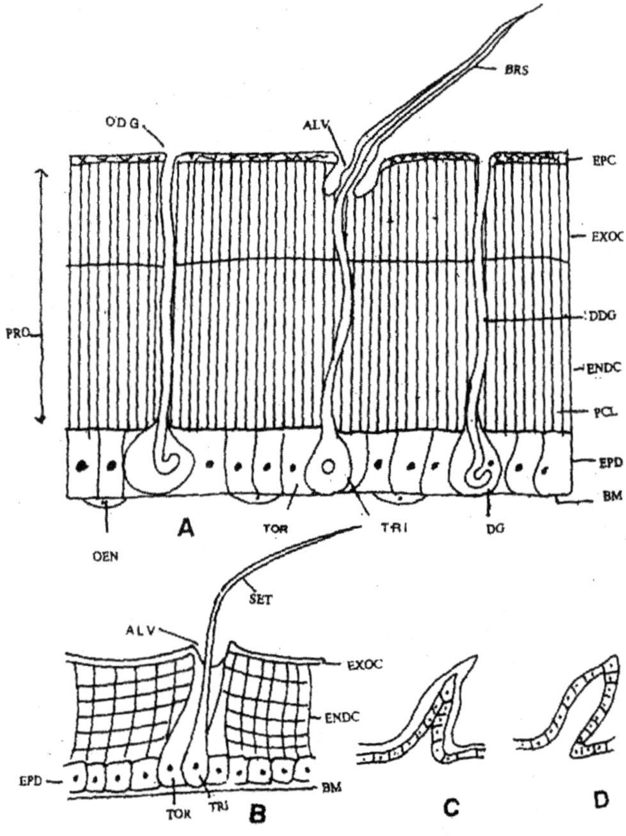

Fig. 1. A. section of typical insect cuticle, B. A tpical seta, C.D. Multi celullar process. *Abbreviations* - ALV - alveolus, BRS - bristle, BM - basement membrane, DG - dermal gland, DDG - duct of dermal gland, EPD - epidermis, EPC - epicuticle, EXOC - exocuticle, ENDC - endocuticle, ODG - opening of dermal gland, OEN - oenocyte, PCL - pore canal, PRO - procuticle, TOR - tormogen cell, TRI - trichogen cell, SET - seta.

THE HEAD

The head is a compact box like structure which supports the mouth parts, antennae and eyes and within it lies the brain. The posterior surface of the head has a large opening, foramen magnum, which leads into the neck and thorax.

Orientation of the Head

According to the position of mouth parts the head may assume one of the three types of structures. If the mouth parts are directed downward and the cranium corresponds in position to the body segments the head is said to be "hypognathous" e.g. Orthoptera, if the cranium is turned upward on the neck so that the mouth parts are directed forward the head becomes "prognathous" e.g. Coleoptera. The third, "opisthognathous" type results from a deflection of the facial region, giving the mouth parts a posterior ventral position e.g. Hemiptera.

Sclerites and sutures (Fig. 2A, B)

In the head, certain areas are strongly sclerotized known as sclerites. The sclerites are separated from each other by narrow lines termed as sutures. The sclerites and sutures represent the skeletal structures that can be recognized well in the insect head. In the special cases a number of structures may be modified or variously changed according to special requirements. In the head of a generalized insect e.g. grasshopper or locust the following structures are clearly recognized *viz.* (1) the frons (2) clypeus (3) labrum (4) vertex (5) occiput (6) genae (7) post genae (8) compound eyes (9) ocelli (10) antennae and (11) tentorium.

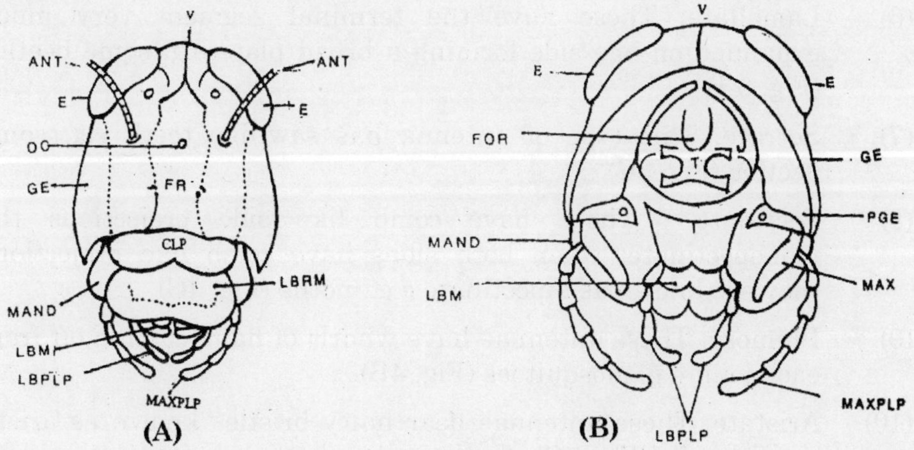

(A) (B)

Fig. 2. A. Frontal view of a grasshopper head, **B.** Posterior view of a grasshopper head. *Abbreviations* - ANT - antenna, CLP - clypeus, E - eye, FR - frons, FOR - foramen magnum, GE - gena, LBM - labium, LBRM - labrum, LBPLP - labial palp, MAND - mandible, MAX - maxilla, MAXPLP - maxillary palp, OC - ocellus, PGE - post gena, T - tentorium.

(A) The Antennae (Figs. 3, 4)

The insect head is provided with a pair of antennae. Each antenna is a slender structure composed of a number of parts. The antenna is made up of three regions viz., the first region called scape, the second pedicel and the remaining part is known as flagellum. They are of different form in different orders of insects. Some are undoubtedly modified for increasing surface area for receptors but many plumose antennae have no more receptors than filiform types. The common types of antennae are the following:

(1) Filiform: These are long, thread like antennae e.g. grasshoppers (Fig. 3E).

(2) Moniliform: They are made up of globose segments and are beaded in appearance e.g. termites (Fig. 3H)

(3) Setaceous: These are bristle like antennae where the size of the segments decreases from the base to the apex region e.g. cockroaches (Fig. 3A)

(4) Clavate: The last segment of the antenna is oval or like a cone e.g. butterflies (Fig. 3B)

(5) Capitate: They form a terminal large knob e.g. some beetles (Fig. 3F)

(6) Lamellate: These have the terminal segment very much expanded on one side forming a broad plate e.g. some beetles (Fig. 3-I)

(7) Serrate: This type of antenna has saw-like teeth e.g. some beetles (Fig. 3C)

(8) Pectinate : These have comb like side projections the monopectinate; when both sides have comb like projections they are known as bipectinate e.g. moths (Fig. 4C)

(9) Plumose: These antennae have whorls of hairs coming off from each joint e.g. mosquitoes (Fig. 4B).

(10) Aristate: These antennae bear hairy bristles known as arista e.g. housefly (Fig. 3G)

(11) Geniculate: It is bent sharply at the distal end of the scape e.g. wasps (Fig. 4A)

(12) Stylate: These antennae have a hairy bristle called style at the end of the last segment e.g. tabanus fly (Fig. 3D)

(13) Flabellate: The anatenna has a special apparatus known as Flabella (Fig. 4D)

(14) Whorled: The antennae having a whorl of bristles at every node (Fig. 4E)

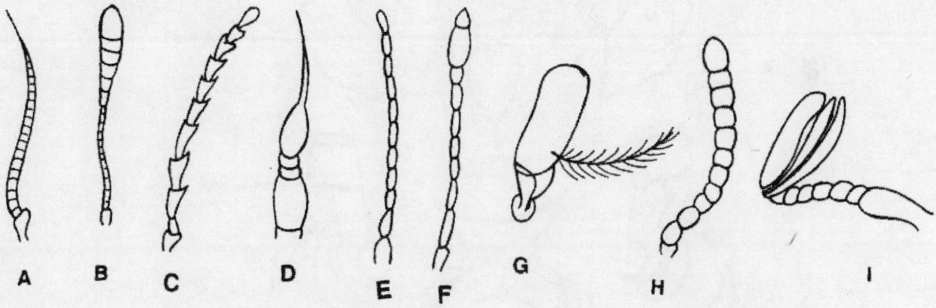

Fig. 3. Types of antennae: **A.** Setaceous, **B.** Clavate, **C.** Serrate, **D.** Stylate, **E.** Filiform, **F.** Capitate, **G.** Aristate, **H.** Moniliform, **I.** Lamellate.

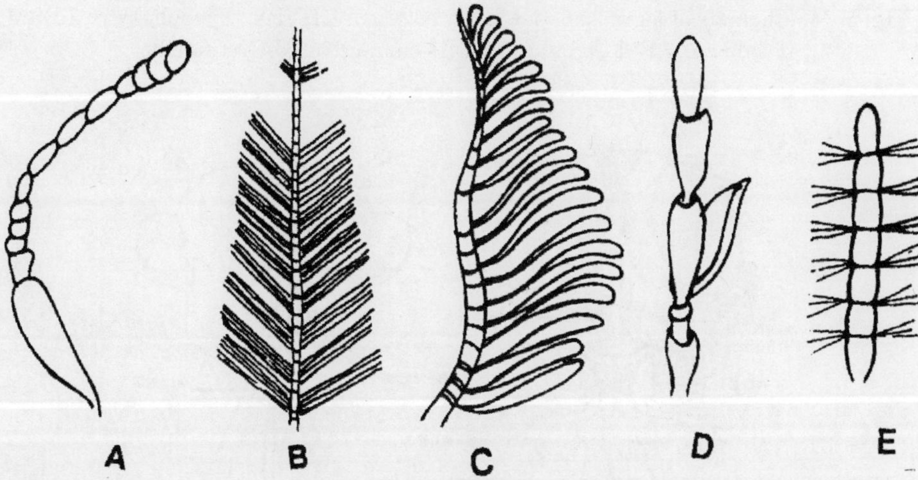

Fig. 4. Types of antennae: **A.** Geniculate, **B.** Plumose, **C.** Pectinate, **D.** Flabellate, **E.** Whorled.

(B) The Mouth Parts (Figs. 5-7)

The appendages of the head which are situated around the mouth and which help the insects in the act of feeding are known as the mouth parts The mouth parts (Fig. 5A) in a generalized insect comprise of the following:

(1) Labrum (2) Epipharynx (3) Mandibles

(4) Maxillae (5) Hypopharynx and (6) Labium

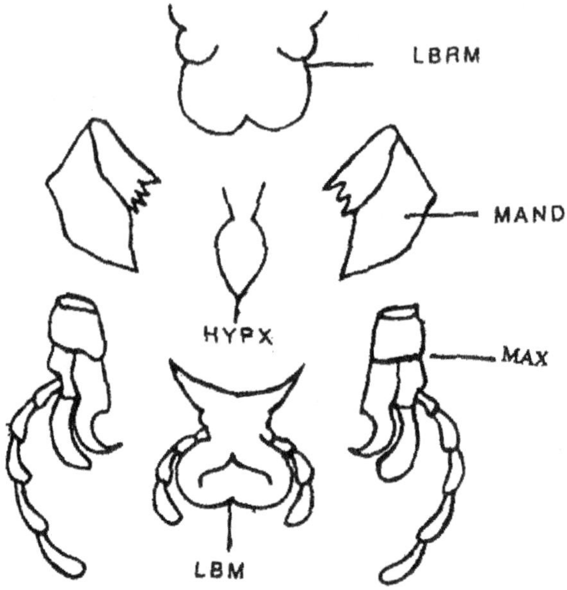

Fig. 5. Mouth parts of a grasshopper - *Abbreviations* - HYPX - hypopharynx, LBRM - labrum, LBM - labium, MAND - mandible, MAX - maxilla

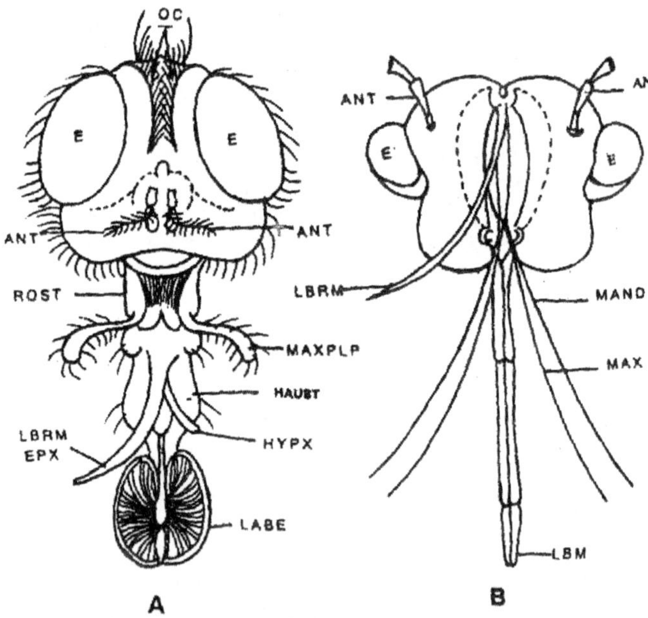

Fig. 6. A. Sponging type of mouth parts, **B.** Piercing and Sucking type of mouth parts *Abbreviations* - ANT - antenna, E - eye, HAUST - haustellum, HYPX - hypopharynx, LABE - labellum, LBRM - labrum, LBM - labium, LBRMEPX - labrum-epipharynx, MAND - mandible, MAX - maxilla, MAXPLP - maxillary palp, OC - ocellus, ROST - rostrum.

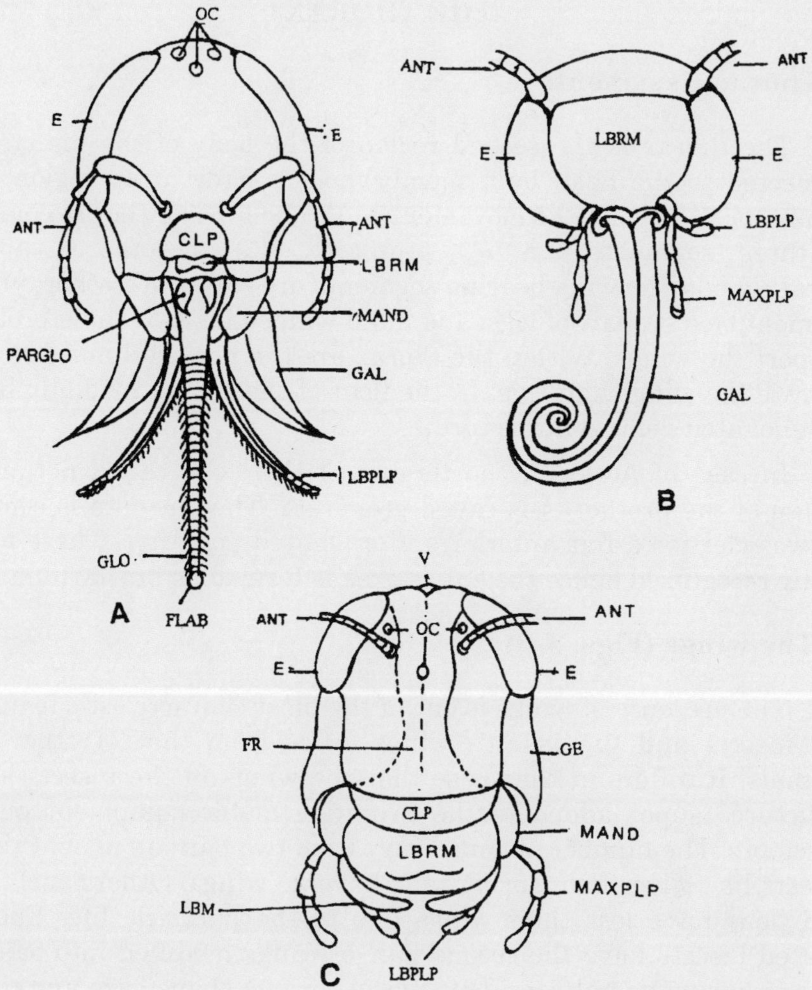

Fig. 7. A. Chewing-lapping type of mouth parts, **B.** Siphoning type of mouth parts, **C.** Chewing type of mouth parts - *Abbreviations* - ANT - antenna, CLP - clypeus, E - eye, FLAB - flabellum, GE - gena, GAL - galea, GLO - glossa, LBM - labium, LBRM - labrum, LBPLP - labial palp, MAND - mandible, OC - ocellus, PARGLO - paraglossa.

Types of mouth parts

Insects feed on plants and animals in a variety of ways and their mouth parts have become modified for these purposes. Here, an account of only the representative of the more important types of mouth parts is given. (1) The piercing and sucking type (Fig. 6, B) (2) chewing type (Fig. 7C) (3) rasping sucking type (4) siphoning type (Fig. 7B) (5) sponging type (Fig. 6A) (6) chewing – lapping type (Fig. 7 A) and (7) degenerate type.

THE THORAX

A. Thoracic segments

The thorax is the second region of the body of the insects. It is connected to the head by a membranous narrow neck region which permits a wide range of movement to the head. The thorax comprises of three segments *viz.* (1) prothorax, (2) mesothorax and (3) metathorax. (The wing bearing segments are called pterothorax). Each segment bears a pair of legs, and in all winged insects, the pterothorax support the wings. Within the thorax are the powerful muscles which operate the wings and legs. In the floor of the thorax medially lies the ganglionated ventral nerve cord.

In case of Acridids, the tergum of thorax is called notum. The notum of the prothorax is called pronotum. The sternum is composed of two sclerites a flat anterior and a pointed posterior. These are not easily recognized hence the entire area is termed as prosternum.

B. The wings (Figs. 8, 9)

The presence of wings is one of the most characteristic features of the insects and the insect's wings differ from that of other flying animals. It differs in the sense that the wings in the insects are the structures super added to the primitive motor equipment on their ancestors. The number of wings vary from two pairs to none. Primitive insects like silver fish, spring tails have no wings. Others such as lice and fleas have lost their wings due to the parasitic life. Flies and related insects have the second pair of wings modified into balancing organs known as halteres. Due to the various shape, size and texture, the classification of insects is based on the characters of the wings.

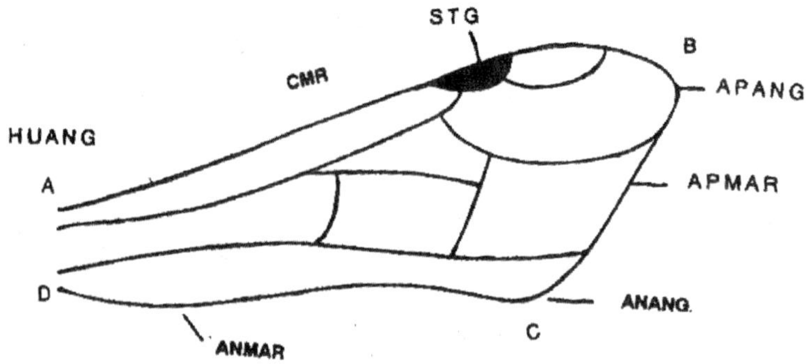

Fig. 8. Wing of a Hymenoptera

Abbreviations - APANG - apical angle, APMAR - apical margin, ANANG - anal angle, ANMAR - anal margin, CMR - costal margin, HUANG - humaral angle, STG - stigma

Pterostigma or stigma (Fig. 8): In many insects near the costal margin of the wing a conspicuous opaque spot is found which is known as the stigma or pterostigma e.g. this pterostigma is found in the forewings of order Plecoptera and most Hymenoptera and in both the pairs of wings of Odonata.

Wing Venation: The mature wing appears as thin membrane with longitudinal thickenings called veins. The arrangement of veins in the wing is termed as venation. Different orders have different types of venation. Much studies have been given to wing venation in order to determine the primitive type characteristics of the common ancestors of all Pterygota, because this may be used in establishing the evolutionary relationships. It was Cosmstock and Needham (1898, 1918) who evolved a method by which the venation of all the orders can be derived. Generalized system of venation was studied at the time of development and it was studied by the tracheae of the wings. The veins passing in the furrows are the concave veins and those of the ridges are termed as the convex veins. Convexity is shown by "+" sign and concavity by "–" sign. The wing venation must have evolved from a wing like the hypothetical form as shown in Figure 9.

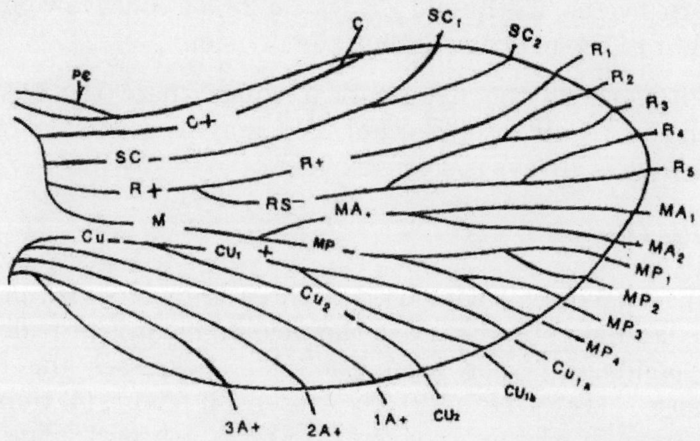

Fig. 9. Hypothetical plan of wing venation showing various veins. *Abbreviations* - 1A+, 2A+, 3A+, branches of anal veins, MA - anterior media, MP - posterior media, MA1, MA2, MP1, MP2, MP3, MP4, - branches of the anterior and posterior media, RS - radial sector, CU1a - anterior cubitus, CU1b - posterior cubitus, PC - pre costa, C - costa, SC - sub costa, R + radials, M - media, CU - cubitus, SC1 - subcosta 1, SC2 - sub costa 2, R1, R2, R3, R4, R5 - branches of radius, CU1, CU2, branches of cubitus, R - radius.

Hypothetical wing venation: (Fig. 9) The name applied to the longitudinal and cross veins and the abbreviations used to designate them are as follows:

1. The precosta (PC) is found in the fossil insects only and it is very small vein.
2. Costa (C+): It is a convex vein running along the margin of the wing and is unbranched.
3. Subcosta (Sc-): It is a concave vein with 2 branches SC1 and SC2.
4. Radial (R+): It is a convex vein with two main branches R1+ passing straight along the margin and Rs- is concave vein passing also along the margin. The Rs branch is divided into 4 branches viz. R2 + R3, R4 + R5, thus there are altogether 5 branches of the radial. The Rs branch is known as radial sector.
5. Media (M-): It is a concave vein and divides into two branches median anterior (MA+) and median posterior (MP-). The former one divides into two branches MA1 and MA2 and MP which is concave one, divides into 4 branches MP1, MP2, MP3 and MP4.
6. Cubitus (Cu-): It is also a concave vein divided into two branches Cu1 + and Cu2-. The Cu1 divides into 2 branches anterior and posterior veins: Cu 1a and Cu 1b. Cu 2- is a concave vein passing straight to the margin without any division.
7. Anals (IA+): In the hypothetical wings there are 3-A, out of which the 1A+ is a convex and 2A ± may be a concave or convex vein. 3A + is always is a convex vein.

C. The Legs (Figs. 10, 11)

The insect thorax is composed of three segments viz., prothorax, mesothorax and metathorax. Each thoracic segment is provided with a pair of appendages – the legs. Each leg, possesses the following movable parts – (1) the coxa, (2) trochanter, (3) femur, (4) tibia and (5) tarsi. Pretarsus is present in some insects as 6th part (Fig. 10). The coxa, femur and tibia are single pieces in each leg while the trochanter is composed of two and the tarsal region 1 – 5 components except in Collembola where the tarsus is fused with the distal end of the tibia and called tibiotarsus. Two claws with a pad – the pulvillus, arolium or empodium in between, are attached in the distal end of the last tarsus and the entire structure is called pretarsus. For different purposes, the various segments of the legs are variously modified and they also bear a number of cuticular structures meant for a definite function.

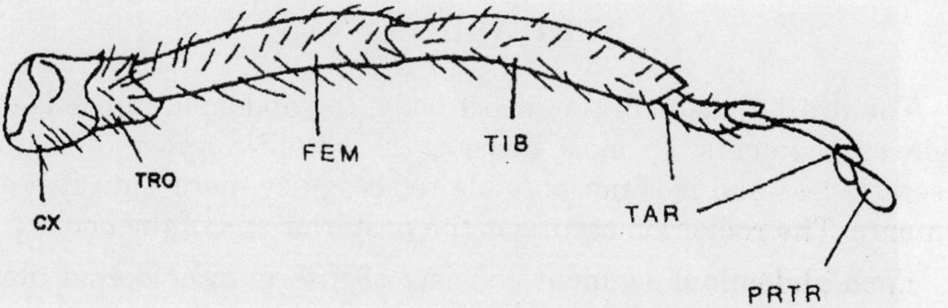

Fig. 10. Structure of an insect leg. *Abbreviations* - CX - coxa, FEM - femur, PRTR - pretarsus, TIB - tibia, TAR - tarsus, TRO - trochanter.

The legs in insects are highly modified and accordingly perform different functions like (1) walking – ambulatorial e.g. bugs, (2) running – cursorial e.g. tiger beetles and cockroaches (3) clinging – scansorial e.g. louse, (4) sound producing – stridulatorial e.g. crickets, (5) hopping – saltatorial – e.g.: grasshoppers (6) digging – fossorial e.g. molecrickets, (7) swimming – natatorial e.g. water beetles (8) grasping – raptorial e.g. mantids etc. (Fig. 11A-H).

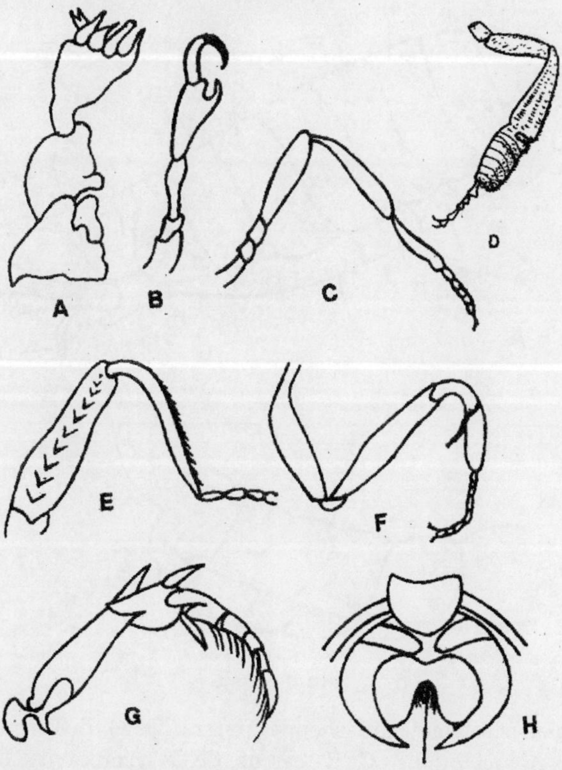

Fig. 11. Different types of legs. **A.** Fossorial, **B.** Scansorial, **C.** Ambulatorial, **D.** Foragial, **E.** Saltatorial, **F.** Raptorial, **G.** Natatorial, **H.** Foot of housefly adapted to walk upside down (From Metcalf and Flint, 1962).

THE ABDOMEN

The third division of the insect body, the abdomen, consists of ten or eleven segments. In most insects, the actual number of segments present is less and it is not possible to recognize more than five or six segments. The reduction occurs at the posterior end of the body.

Each abdominal segment consists of a tergum or dorsal plate, a sternum or ventral plate, lateral areas of membrane connecting tergum and sternum and a spiracle on each side usually situated in the lateral membrane. In Acrididae in the first abdominal segment is located a large tympanic membrane that covers the ear which is the characteristic feature of the family.

In most insects, the 8th and 9th abdominal segments are greatly modified to form reproductive appendages or the external genitalia. The 9th abdominal segment in the male insects and the 8th and 9th in the female insects form the genital apparatus (Fig. 12 AB).

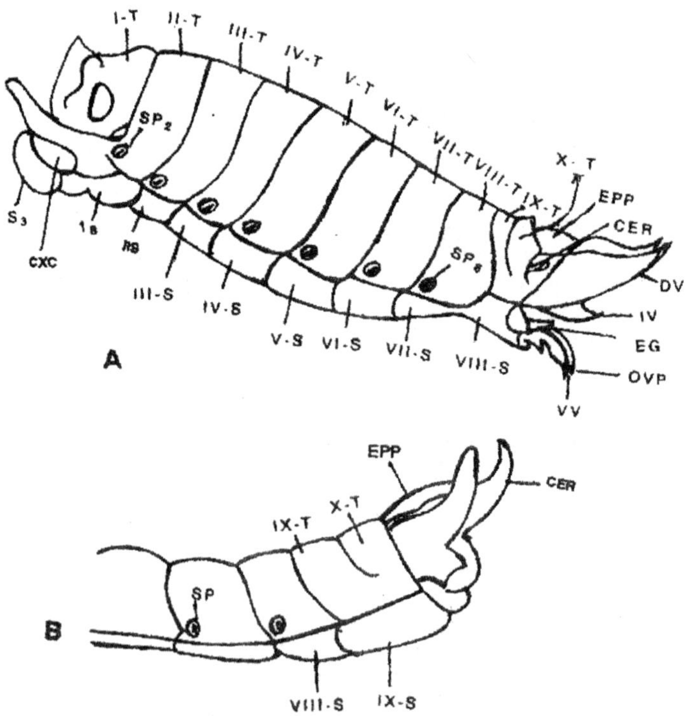

Fig. 12. A. Abdomen of a female grasshopper (lateral view), **B.** Terminal segment of a male grasshopper. *Abbreviations* - CER - cercus, CXC - coxal cavity, DV - dorsal valve, EG - egg guide, EPP - epiproct, IV - inner valve, OVP - ovipositor, S3 - sternum of 3rd thoracic segment, IS-IX S 1 to 9th abdominal sterna, SP2 - SP8 2 to 8th spiracles, I-T to XI-T - 1st to 11th terga, SP- Spiracle, VV-ventral valve.

Male genitalia

As stated above the 9th abdominal segment is the genital segment in male insects in which also lies the male genital opening (Fig. 12 B).

In a typical male Pterygote insect, the complete external genitalia are composed of (1) penis, (2) a pair of parameres or the inner processes and (iii) a pair of claspers that serves to grasp the female when mating and between which lies the aedeagus, within the aedeagus lies the penis. The penis is formed by the fusion of paried outgrowths and it bears the male gonopore. The original paired condition is found only in Ephemeroptera and in certain Dermaptera. The inner processes or parameres are present in several orders but it is not certain that they are homologous in all cases. The claspers or gonopods are derived from the appendages of the 9th segment (Fig. 12B).

Female genitalia

In all Pterygote insects the ovipositor is composed of three pairs of valves (or valvulae) *viz.* (1) anterior or ventral, (2) posterior or inner and (3) lateral or dorsal. The first pair, anterior or ventral valves, arise from the valvifer – 1 of the 8th segment. The 2nd, posterior or inner valves are the median outgrowths of the valvifer – 2 of the 9th segment and the 3rd pair, lateral or dorsal valves arise from the valvifer 3 of the 9th segment. These three pairs of valves (valvulae) are variously modified in several insects (Fig. 12A).

THE DIGESTIVE SYSTEM

The Alimentary Canal

The alimentary canal in insects is a tubular structure of variable length arising as an infolded part of the body wall. The structure of the wall of the alimentary canal is therefore essentially the same as that of the integument. It is this part of insect anatomy which discharges the functions of digestion, absorption and egestion of the food materials. Associated with the alimentary canal are the organs of ingestion mainly the mouth parts and other glands such as salivary glands, and hepatic caecae which pour their secretions into it. The entire alimentary canal extends from mouth to the anus in insects. This tubular structure is often longer than the body of the insect and therefore variously convoluted upon itself. In fact, the length of the alimentary canal and the complexity of its structure depends on the food habits of the insects. The alimentary canal in insects is divisible into three primary region on the basis of embryonic origin viz. 1. Stomodaeum or foregut, 2. Mesenteron or midgut or ventriculus and 3. Proctodaeum or hindgut. At the junction of foregut and midgut are present the enteric caecae or hepatic caecae which pour their secretions into the mid gut (Fig. 13AB).

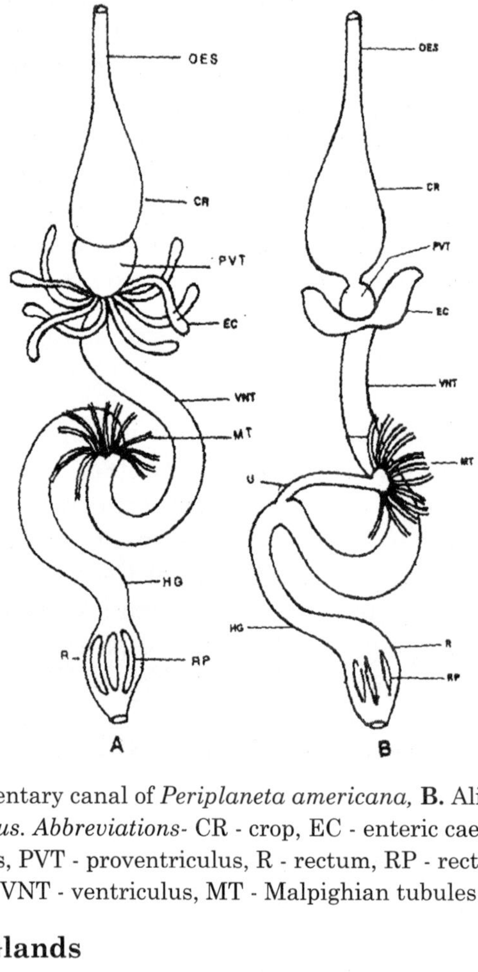

Fig. 13. A. Alimentary canal of *Periplaneta americana,* **B.** Alimentary canal of *Gryllodes sigillatus. Abbreviations-* CR - crop, EC - enteric caeca, HG - hind gut, OES - oesophagus, PVT - proventriculus, R - rectum, RP - rectal pad, U - ureter, VNT - ventriculus, MT - Malpighian tubules.

The Cephalic Glands

In the majority of insects there are present cephalic glands or head glands (Fig. 14AB) of diverse nature that open either in the buccal cavity or in the pharynx. Some of them secrete a slimy substance i.e. saliva while in others they have also a number of digestive enzymes in their products. There are mainly four types of cephalic glands: viz. (1) *Mandibular glands.* These are small glands that open near to the base of the mandibles in Apterygota, Orthoptera, Isoptera, Dictyoptera, Coleoptera, Trichoptera and Hymenoptera. In *Apis* the glands are larger in queens than in the workers and are very small in drones. In the queen they produce the pheromones concerned with colony control while in workers, they produce some saliva and serve to soften the cocoon at the time of emergence. (2) *Maxillary glands.* These are present in some primitive group of insects e.g. Collembola, Protura, in the larvae of Neuroptera and certain Trichoptera. They open at the base of maxillae and lubricate the mouth parts. (3) *Labial glands.* The labial glands are commonly known as salivary glands and are paired structures, generally situated in the thorax on either side of stomodaeum. They are present in majority of the insects and assume a great variety of form and structure. (4) *Pharyngeal glands.* The pharyngeal glands occur in Hymenoptera and are well developed in worker honeybees. They are vestigeal in the queen bee and absent in male. The glands open at the base of the hypopharynx by separate ducts; they produce brood food with which young larvae are fed and which plays some role in caste determination. It also produces diet of laying queen and drone.

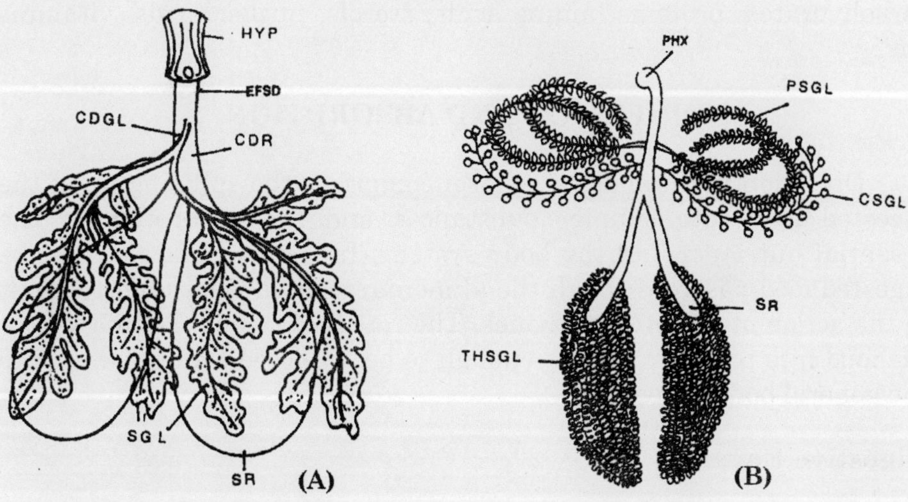

Fig. 14. A. Salivary glands of a cockroach, **B.** Salivary glands of a Honey bee. *Abbreviations* - CDR - common duct of reservoir, CDGL - common duct of gland, EFSD - efferent salivary duct, HYP - hypopharynx, SR - salivary receptacle or reservoir, SGL - salivary glands, PHX - Pharynx, PSGL - pharyngeal salivary gland, CSGL - cephalic salivary gland, THSGL - thoracic salivary gland.

NUTRITION

Nutrition is the study of the food requirements of organisms, in this case, insects (Patton, 1963). House (1961, 1962) remarked that insect nutrition concerns more than mere dietetics. In its broadest sense, it involves many metabolic processes. It is a bond between physiologic and ecologic phenomena associated with natural selection and competition for food.

Nutritional requirements

The food constituents are classified into two major groups essential and nonessential. The essential designates to those components that must be included in the diet, because they cannot be synthesized by the metabolic system of the insects. The nonessential indicates food materials that may have to be consumed to produce energy and that can be converted into a form in which they can be utilized through metabolic processes. The examples of essentials are vitamins, amino acids and some mineral salts. Although carbohydrates, proteins and fats are individually non-essential, but the animal must eat at least one of these nonessential to maintain life. In order to grow and reproduce, insects like most other animals require in their food an adequate quantity of water, minerals, fats, carbohydrates, proteins, amino acids, sterols, nucleic acids, vitamins etc.

DIGESTION AND ABSORPTION

Digestion in animals in general comprises the splitting up of the ingested food into simpler substances and the absorption of the essential nutrients into the body systems In insects, likewise, as the ingested food passes through the alimentary canal, it is subjected first to the action of digestive enzymes. The result of this is the analysis of the food into products simple enough to be absorbed straight way into the general body system.

Digestive Enzymes

The occurrence of digestive enzymes in an insect is governed by a variety of factors. One of the chief factors is the ability of the epithelial cells lining the alimentary canal to secrete these digestive juices. It is generally believed that the mid gut is the chief site of digestion in insects, since this region shows the maximum enzymatic activity.

Another factor which governs enzyme secretion and occurrence, is the presence or absence of food inside the alimentary canal. It has been suggested that enzyme secretion is continuous in a majority of insects and to those larvae which are immersed in or always close to their food and is intermittent in many carnivorous forms and other insects which feed at definite intervals. In the latter group of insects, however such secretogogues or other stimuli may be present which activate enzyme secretion.

The bulk of food material available to insects is constituted by the three main classes of foodstuffs viz. - carbohydrate, protein and fat in proportions depending on the diet of the different species. Therefore, a corresponding diversity of enzymes capable of hydrolyzing these substances will be required by most of insects. Depending on the nature and amount of the food material, the carbohydrases, proteinases and lipases are secreted in large or small concentration. In addition, a number of insects feed on chitin, keratin, beeswax, lichinose etc. and for their digestion, specific enzymes are developed only in those specific insects.

Absorption

As soon as the food splits up into simpler products the next step is the study of the insect gut physiology and ability of these organs to absorb their products. Evidence for absorption studies depends mostly upon the accumulation of the absorbed quantities of materials in the cells. But there is no doubt that many, materials are passed through the epithelial cells of the gut without accumulation. Thus it is clear that absorption is an active and often a highly selective process, certain groups of cells being specially concerned with the uptake of specific material from the lumen.

THE CIRCULATORY SYSTEM

The circulatory system in insects is of open or lacunar type. The blood circulates in the haemocoel. In insects this system plays a very insignificant role in respiration. There is only one palsatile vessel.

The diaphragms and sinuses

The body cavity or haemocoel is divided into three sinuses by means of diaphragms (Fig. 15A). (1) The dorsal diaphragm extends across the abdominal cavity above the alimentary canal. The blood space thus formed is known as dorsal or pericardial sinus. It encloses the heart. (2) The ventral diaphragm stretches across the abdominal cavity just above the ventral nerve cord. The haemocoel thus formed is known as ventral or perineural sinus. (3) Between the dorsal and ventral sinuses is a large cavity known as visceral or perivisceral sinus which contains the main internal organs.

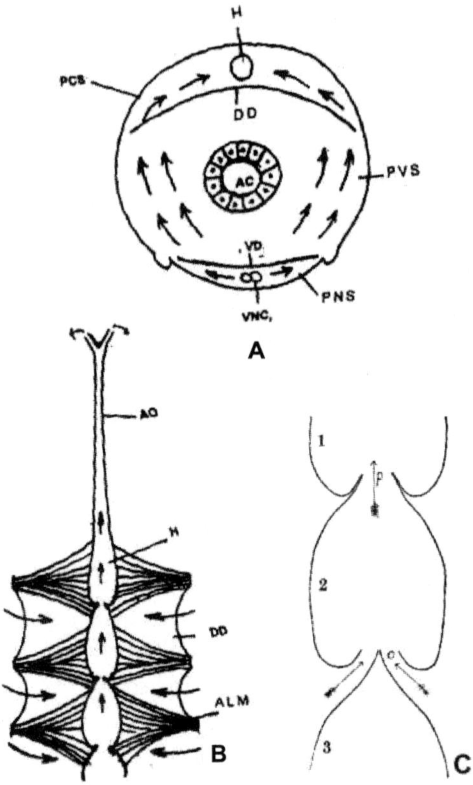

Fig. 15. A. Transverse section of abdomen, **B.** Dorsal vessel showing a few chambers of the heart, alary muscles and dorsal diaphragm, **C.** Valves of the heart. 1,2,3- chambers of the heart; 1,2-at the moment of systole; 2,3- at the moment of diastole (Fig. C from Imms, 1963, 1964) *Abbreviations* - AC - alimentary canal, AO - aorta, ALM - alary muscles, DD - dorsal diaphragm, H - heart, O - ostium, P- intraventricular passage, PCS - pricardial sinus, PNS - perineural sinus, PVS - peri visceral sinus, VD - ventral diaphragm, VNC - ventral nerve cord.

Pairs of alary muscles which arise from the terga spread out fan like over the surface of the dorsal diaphragm. Thus beneath the alary muscles lies the dorsal diaphragm. The alary muscles are attached to the body wall along the lateral parts of the dorsum. They are believed to function more or less as elastic sheets which bring about contraction and expansion of the heart.

Dorsal vessel

The dorsal vessel (Fig. 15B,C) runs along the mid dorsal line just beneath the integument for the whole length of the body. It is a continuous tube whose posterior end is closed and is always open at the anterior or cephalic extremity. The dorsal vessel is divided into two regions : a posterior heart and an anterior aorta. The heart is perforated while the aorta is an unperforated tube. The wall of dorsal vessel in the heart and aorta is contractile.

COMPOSITION OF BLOOD

The blood or haemolymph in insects is composed of liquid plasma or serum and blood cells or haemocytes. It makes up 15-75% of the volume of the insect and the plasma contains about 85% water. It is a colourless fluid but occasionally it appears with a greenish, yellowish or brownish tinge. The plasma is slightly acidic in insects in general. The pH of insect blood varies considerably depending upon the age of the insect, it may be slightly acidic or alkaline as in the case of *Bombyx mori.* The chemical composition of the insect blood reveals the presence of proteins, lipids, sugars (glucose in case of honeybee and trehalose in others), uric acid, a number of pigments, organic acids, are found in the insect haemolymph. Some 19-free aminoacids have been reported. The plasma of *Periplaneta* has more sodium than potassium but this is not true of all insects, the ratio of sodium and potassium varies from insect to insect. On the basis of sodium and potassium ratio the blood has been classified into two types. In Odonata the concentration of sodium is greater than the concentration of potassium and magnesium, in Lepidoptera, Coleoptera etc. the plasma has a low concentration of sodium and potassium, and magnesium have a high concentration.

THE EXCRETORY SYSTEM

In insects, the function of excretion is chiefly concerned with the elimination of several types of substances such as mineral salts and water, and products of the metabolism of nitrogenous substances, sulphur and phosphorus; certain complex compounds that are produced incidentally as a result of certain accidents or as a by product in case of certain chemical changes and the acids and alkalies. Thus, it is a process by which the internal environment of the insects is regulated at a constant level by removing surplus and injurious materials.

In general, the organs responsible for the elimination of the substances, mentioned above are known as excretory organs. There are several types of excretory organs in insects chief of them are Malpighian tubules (Fig. 16A-D).

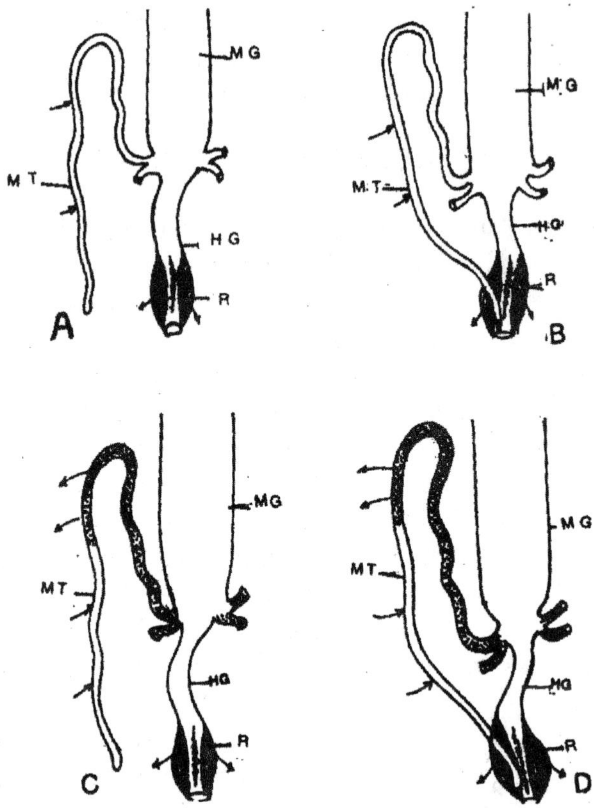

Fig. 16. A. Free floating Malpighian tubules as found in Orthoptera, **B.** Re-association of Malpighian tubules as found in Coleoptera, **C.** Malpighian tubules as found in Hemiptera, **D.**Re-association of Malpighian tubules as found in Lepidoptera. *Abbreviations* - HG - hindgut, MG - midgut, MT - Malpighian tubule, R - rectum.

THE RESPIRATORY SYSTEM

In insects respiration takes place through a system of internal tubes known as tracheae, the finer branches of which convey air directly to the tissues termed tracheoles. The tracheae open along the side of the body by a series of apertures, termed spiracles. (Fig. 17AB)

General plan of tracheation

The general plan of body tracheation (Fig. 17A) in insects is as follows:

From each spiracle on either side is given off a short tracheal branch known as the spiracular trachea which joins the main tracheal trunk running along the full length of the body. The two main tracheal trunks are known as the lateral longitudinal trunks and they form usually the largest trachea in insects. Spiracular trachea or the lateral longitudinal trunk of each side gives off three branches viz. (1) a dorsal trachea entering the dorsal musculature of the body wall and to the dorsal blood vessel, (2) a ventral trachea supplying the ventral musculature and the ventral nerve cord and sending a branch to the leg in the leg bearing segment and (3) a median visceral trachea having its principal ramifications on the walls of the alimentary canal with branches to the fat body and in appropriate segments to the reproductive organs. In some insects, the dorsal, ventral and visceral trachea of successive segment become continuous to form the respective longitudinal trunks. Often the dorsal trachea of one side of a segment is connected by a commissure transversely to its fellow of the opposite side in the same segment. This is the dorsal tracheal commissure. There is a similar ventral tracheal commissure also in certain insects.

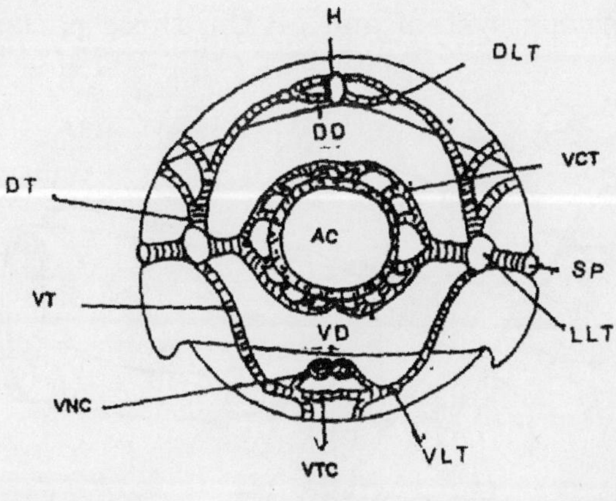

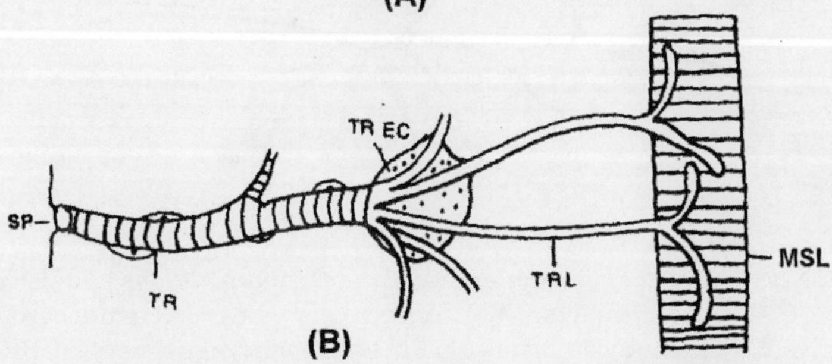

Fig. 17. A. Diagrammatic cross section of abdomen showing main tracheae and transverse trunks. **B.** Diagram showing open rachea of a generalised insect. *Abbreviations* - AC - alimentary canal, DD - dorsal diaphragm, DT - dorsal trachea, SP - spiracle, TR - trachea, TRL - tracheole, TREC - tracheal end cell, MSL - muscles, VD - ventral diaphragm, VT - ventral trachea, VCT visceral trachea, VNC - ventral nerve cord, VTC - ventral tracheal commissure, VLT - ventro lateral trachea.

THE NERVOUS SYSTEM

The nervous system is composed of nerve cells or neurons which are ectodermal in origin. The nerve cell consists of a cell body containing the nucleus and long cytoplasmic projections or axons which extend to make contact with other neurons. Frequently, the axons have branches, and collaterals and end in a terminal arborization. Nerve impulses are transmitted from one neuron to the next via axons.

The nerve cells are of the three types *viz.,* monopolar, bipolar and multipolar. Generally the nerve cells of the insects are monopolar having only one axon. The peripheral sense cells are bipolar. The hypocerebral ganglion and frontal ganglion have some cells with more than two axons and represent the multipolar cells.

The nervous system can be distinguished into three distinct parts *viz.,* (1) central nervous system (2) sympathetic nervous system and (3) peripheral nervous system and all the three parts are connected with each other.

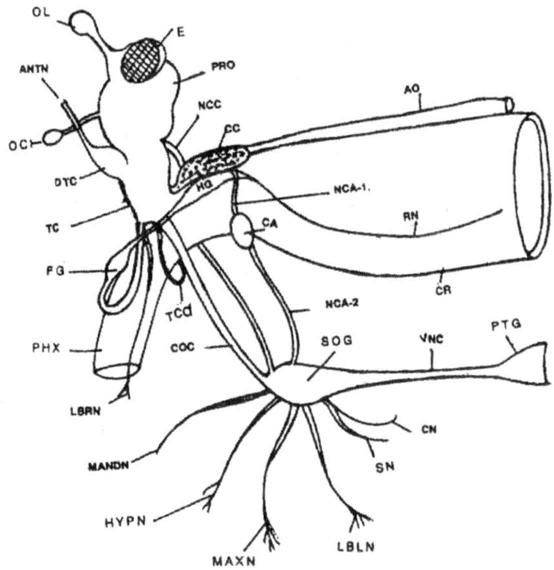

Fig. 18. Nervous system of a generalised insect. *Abbreviations* - ANTN - antennal nerve, C - corpus cardiacum, CN - cervical nerve, COC - circum oesophageal connective, DTC - deutocerebrum, HYPN - hypopharyngeal nerve, LBRN - labral nerve, LBLN - labial nerve, MANDN - mandibular nerve, MAXN - maxillary nerve, OC - ocellus, OL - optic lobe, PRO - protocerebrum, SN - salivary nerve, TC - tritocerebrum, TCC - tritocerebral commissure, VNC - ventral nerve cord, E - eye, CA - corpus allatum, AO - aorta, NCC - nervi corporis cardiaci, NCA1-NCA2-nervi corporis allati 1 and 2, SOG - suboesophageal ganglion, PTG - prothoracic ganglion, PHX - pharynx, FG - frontal ganglion, RN - recurrent nerve, CR - crop, HG - hypocerebral ganglion.

(1) Central Nervous System

It consists of an anterior median brain, located in the head above the alimentary canal, suboesophageal ganglion below the oesophagus and the ventral ganglionated nerve cord situated in the mid ventral part of the body, in which the ganglia are united lengthwise by interganglionic connections (Fig. 18).

(2) Sympathetic nervous system

It can be divided into (1) stomodaeal sympathetic nervous system, (2) ventral sympathetic nervous system and (3) caudal sympathetic nervous system.

i. Stomodaeal sympathetic nervous system

It lies on the dorsal surface of the foregut and consists of frontal, hypocerebral and visceral ganglia and recurrent nerve. It is connected with the brain directly through nerves and innervates the foregut, mid gut and other tissues. Account of corpora cardiaca and corpora allata is given with endocrine organs (see page 32).

ii. Ventral Sympathetic nervous system

This system consists of a pair of transverse nerves lying in association with each nerve cord ganglion. There is a median longitudinal nerve which connects the transverse nerve with the preceding ganglion. These transverse nerves innervate the spiracles of their segment.

iii. Caudal sympathetic nervous system

From the last ganglion of the nerve cord in the abdominal region, nerves are given to the reproductive organs, posterior portion of the alimentary canal, and Malpighian tubules which constitute the caudal sympathetic nervous system.

(3) Peripheral Nervous system

It consists of all the nerves radiating from the ganglia of the central and sympathetic nervous systems.

THE PHOTORECEPTOR AND PHOTOGENIC ORGANS

(A) Photoreceptor organs

The photoreceptors of insects are the simple eyes or ocelli and the compound eyes. In an ocellus, only one corneal lens is present whereas in the compound eye the cornea is divided into numerous separate facets. Thus, compound eyes are composed of a number of simple eyes or ommatidia, each ommatidium is the unit structure of the eye.

Ocelli : There are three simple eyes or ocelli which are located between the compound eyes and the frontal region of the head capsule. In each ocellus, there is a single cornea or lens which is formed by the transparent cuticle. It is secreted by the underlying corneagen cells. Beneath the corneagen cells, lie the photo sensitive retina or retinulae. The retina is formed by the retinal cells. The apposed rhabodomeres of the retinal cells form a central or axial rhabdome. Slight variations are found in the ocelli from this basic pattern in different insect orders.

Compound eyes: There are two compound eyes, each lies on either side of the head capsule. Each compound eye is composed of identical units, the ommatidia. The number of ommatidia varies considerably from species to species which may be from one to several thousands e.g. in the worker ant *Ponera*, each eye has only one ommatidium, in the housefly *Musca* each eye consists of about 4,000 ommatidia and in the dragon flies each eye has some 10,000 to 28,000 ommatidia. Various parts of the compound eye are shown in Figures 19AB.

(B) Photogenic organs

There are some insects which are self luminous or light producing i.e. photogenic. The luminous properties possessed by these insects are because (a) some possess special photogenic organs, (b) some other possess this property due to the presence of light producing bacteria (c) in some cases due to ingestion of the luminous food.

Light is emitted when a substance known as "Luciferin" is oxidized by free oxygen in the presence of water and an enzyme Luciferase. The amount of heat produced in the emission of light is very little and energy for this change is got through preliminary reaction evolving ATP. The regulation of flashing in complex organ depends on a nervously controlled organ of a tracheal end cell which admits oxygen to the photogenic tissues.

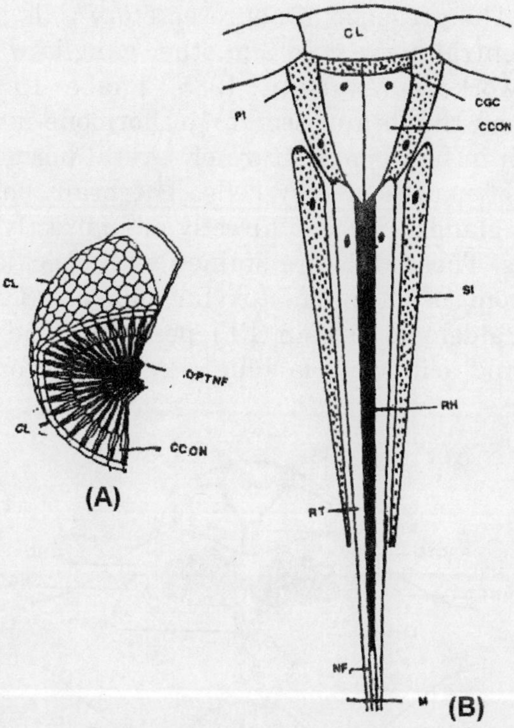

Fig. 19. A. Structure of a compound eye, **B.** Structure of an ommatidium. *Abbreviations* - CGC - corneagen cell, CCON - crystalline cone, CL - corneal lens, M - membrane, NF - nerve fibre, PI - primary iris, RH - rhabdome, RT - retina, SI - secondary iris, OPTNF - optic nerve fibre.

It is supposed that the light helps the insect in the attraction of sexes for mating. In *Photinus* the female attracts the male by giving the flash at an interval of about 2 seconds.

THE ENDOCRINE ORGANS

The endocrine organs of insects are of two types: the neurosecretory cells found in the brain and various ganglia of the nervous system and the endocrine glands like-corpora cardiaca, corpora allata and ecdysial glands (Fig. 20A,B). In some insects the cephalic aorta is also associated with the endocrine organs. (For more details see : Awasthi, 2005, 2009).

1. The Neurosecretory Cells

The neurosecretory cells (Fig. 20A) are the glandular nerve cells which show cytological evidence of secretory activity. The product of these neurosecretory cells is called neurosecretory material (NSM) or neurosecretion. The presence of neurosecretory cells in the brain, the ganglia of the ventral nerve cord and other ganglia was demonstrated by numerous workers (Awasthi, 1985, Raabe 1989). The neurosecretory cells of the brain secrete a hormone known as "brain hormone," which in fact is a mixture of several hormones secreted by different types of neurosecretory cells. The brain hormone activates other endocrine glands and also directly or indirectly controls almost all life processes. These cells are stained with specific neurosecretory stains like chrom alum haematoxylin phloxin (CAHP), aldehyde fuchsin (AF) paraldehyde fuchsin (PF), performic acid resorcin fuchsin (PARF), performic acid victoria blue (PAVB), performic acid alcian blue (PAAB) etc.

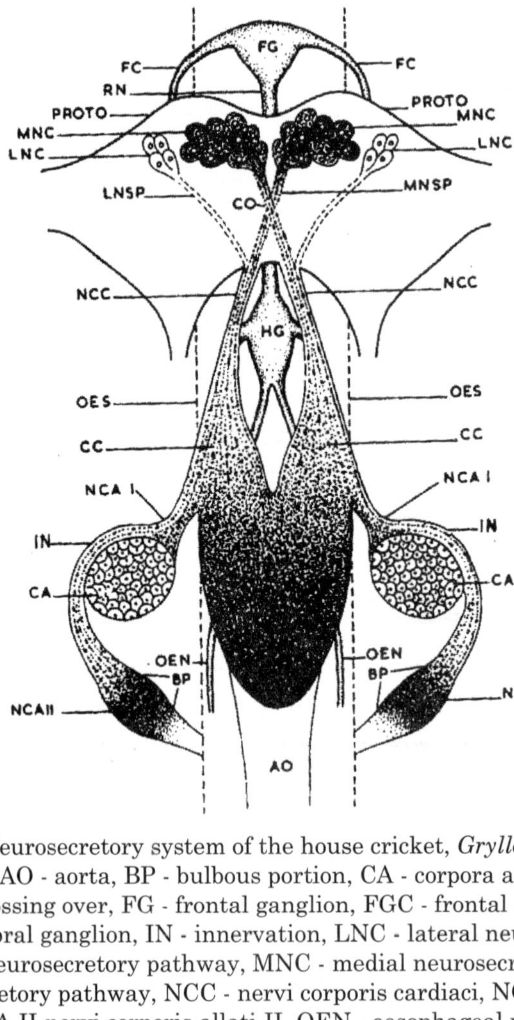

Fig. 20A. Neurosecretory system of the house cricket, *Gryllodes sigillatus*. *Abbreviations* - AO - aorta, BP - bulbous portion, CA - corpora allata, CC - corpora cardiaca, CO - crossing over, FG - frontal ganglion, FGC - frontal ganglion connective, HG - hypocerebral ganglion, IN - innervation, LNC - lateral neurosecretory cells, LNSP - lateral neurosecretory pathway, MNC - medial neurosecretory cells, MNSP - medial neurosecretory pathway, NCC - nervi corporis cardiaci, NCA-I - nervi corporis allati-1, NCA-II-nervi corporis allati-II, OEN - oesophageal nerve, PROTO-protocerebrum, RN - recurrent nerve (From Awasthi, 1969).

With CAHP the A-cells stain blue black and the B-cells pink whereas with PF, AF or PARF the A-cells stain purple, purple red or brick red and the B-cells green. Likewise C and D cells are also distinguished. It is probable that the visible secretion is only a carrier protein (neurophysin) to which the hormone molecule is attached. Thus carrier substance which is the stainable part acts as a marker for the hormone. When hormone is released it becomes separated from the carrier and is then free to enter the haemolymph.

Under the electron microscope also several types of neurosecretory cells have been distinguished on the basis of the size of granules and electron density. The A-cells always contain membrane bound neurosecretory granules 1000 to 4000 Å in diameter and the matrix is electron dense. The granules in B-cells and other types of cells are much smaller in diameter and the matrix may be less electron dense, electron transparent or electron lucent.

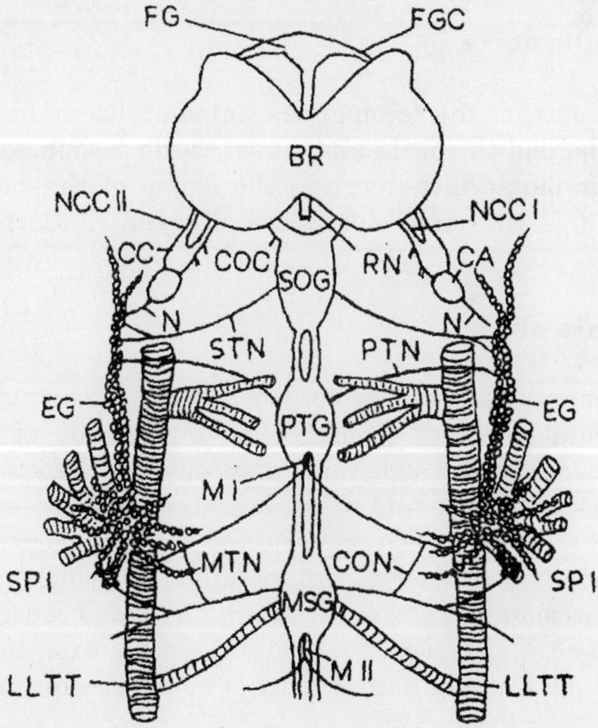

Fig. 20B. Endocrine organs of *Amsacta collaris. Abbreviations* - BR - brain, CC - corpora cardiaca, CA- corpora allata, CON-connective, EG - ecdysial gland, FG - frontal ganglion, FGC - frontal ganglion connective, LLTT - lateral longitudinal tracheal trunk, M-I-medial - I, M-II- medial II, MTN - metathoracic transverse nerve, MSG - mesothoracic ganglion, N - nerve, NCC-I-nervi corporis cardiaci-I, NCC-II-nervi corporis cardiaci-II, PTG - prothoracic ganglion, PTN - prothoracic transverse nerve, SOG - suboesphageal ganglion (From Awasthi and Singh, 1983).

2. The Corpora Cardica

In most of the insect orders the corpora cardiaca (Fig. 20AB) perform the function of a principal neurohaemal organ. The corpora cardiaca are a pair of oval, spherical, spindle-shaped, elongate or U-shaped neuroglandular bodies, lying behind the brain on the dorsal part of the foregut and are associated with the cephalic aorta. They are absent in Collembola. The corpora cardiaca receive the neurosecretory axons of cerebral origin *via* neurosecretory pathways and nervi corporis cardiaci. The cardiac bodies are fused in some insects and separate in others. In some insects the corpora cardiaca are differentiated into two distinct regions, one region is composed of neurosecretory axons and axon terminals of cerebral neurosecretory cells and the other glandular part containing intrinsic secretory cells, but in many insects no such distinction is possible.

3. The Cephalic aorta

Cephalic aorta is the second principal neurohaemal organ. This is also situated behind the brain and dorsal to the oesophagus and has a wider anterior part which receives the axons of the nervi corporis cardiaci - I. It is the main neurohaemal organ in Dermaptera and Heteroptera.

4. The Corpora allata

The corpora allata are non-nervous, epithelioid, oval, spherical or leaf like glandular bodies which lie on either side of the corpora cardiaca. They are paired structures in most of the insects but in some insects e.g. Dermaptera and Gymnoceratan Heteroptera they are unpaired medial bodies. Each is connected with the corpus cardiacum of the same side by the nervi corporis allati - I which carries fibres from the neurosecretory cells of the brain. The gland consists of parenchymal cells and is innervated by neurosecretory axon-terminals and preterminals containing different types of neurosecsetory granules.

Four types of corpora allata have been recognized in insects viz. (1) The vesicular type (2) The pseudolymphoid type (3) The microcellular type and (4) The macrocellular type.

The corpora allata produce juvenile hormone which controls metamorphosis and yolk deposition in the eggs. A number of other functions like tanning of the cuticle, green-brown polymorphism in locust, mating behaviour in social insects, etc. are performed by the juvenile hormone. Since in young stages it maintains larval characters and inhibits adult differentiation, it is also called neotenin or juvenile hormone.

5. The Ecdysial Glands

The ecdysial glands (Fig. 20B) are also known as prothoracic glands or thoracic glands. They are paired, glandular structures, located in the ventrolateral areas of the prothorax extending into the mesothorax in some species and sometimes partially fused in the median line. They are often closely associated with the lateral longitudinal tracheal trunks near the thoracic spiracles (Fig. 20B) In each larval instar there is a distinct cyclicity in the secretory activity. The greatest size and secretory activity of the ecdysial glands is reached in the last larval instar. The ecdysial glands degenerate in majority of the insects after the final moult but they remain preserved in *Locusta* and *Amsacta*. These glands produce ecdysone or moulting hormone.

6. The Weismann's Ring

It is also known as ring gland. It is found in the larvae of Cyclorrhaphous Diptera like house fly, flesh fly and blow fly larvae. The basal part of the ring is composed of fused corpus cardiacum and hypocerebral ganglion, the sides are formed by the cells of the ecdysial glands and the upper part is formed by the corpus allatum. The central part of the ring is occupied by the cephalic aorta. Thus all the components of the retrocerebral endocrine glands participate in the formation of the ring gland. It is connected to the brain by a pair of nerves and is also richly tracheated.

HORMONAL CONTROL OF METAMORPHOSIS

Insect metamorphosis is controlled by a variety of hormones of which the most important are (1) brain hormone (BH), (2) ecdysone or moulting hormone (MH) and (3) juvenile hormone (JH).

The brain hormone secreted by the neurosecretory cells of the brain stimulates the prothoracic glands to secrete ecdysone or moulting hormone, but moulting occures only when the juvenile hormone is either absent or present in very small amounts. In the last larval instar of Holometabola the corpora allata becom almost inactive, hence a low concentration of JH is present causing larva to moult into a pupa. Similarly during later part of the pupa or late pupa, again the corpora allata become inactive and in total absence of JH the pupa moults into an adult insect. In Hemimetabolous insects also the same phenomena is repeated i.e. the corpora allata become inactive in the last instar nymphs and in the complete absence of JH imaginal ecdysis takes place. Thus metamorphosis in insects is controlled by a delicate balance in the timing of secretion and concentration of two hormones viz. JH and MH secreted by the corpora allata and ecdysial or prothoracic glands. The total absence of JH is the most significant factor for imaginal ecdysis in both, Hemimetabolous and Holometabolous insects.

If much JH is present then larva will moult into a larva, if small amount of JH is present then pupation takes place and in the total absence of JH adult emergence occurs. And if the balance of hormones is disturbed then monsters, intermediate between larva and pupa or between pupa and adult are produced and the phenomena is termed as prothetely or metathetely.

Almost all the life processes are controlled by the insect hormones, a comprehensive account of which is given by Raabe (1989) and Awasthi (2005).

THE PHEROMONES

The social attraction of an insect colony appears to be maintained primarily through the utilization of a series of communication chemicals which function to transmit specific bits of information to the members of a single population. These chemicals are termed as "pheromones". The term pheromone was coined by Karlson and Butenandt (1959). According to Karlson and Butenandt (1959) the pheromones are chemicals which are secreted to the outside by animals, in this case insects, and which if passed to another individual of the same species, cause it to respond in a particular manner. These are produced in several specialized tissues or glands called exocrine glands, which evacuate their contents into the environment in response to specific stimuli. Social insects such as hymenopterans and isopterans, possess many types of exocrine glands. Many aspects of insect biology and behaviour are controlled by these chemicals i.e. the pheromones. The pheromones which have been identified are sex pheromones, trail marking pheromones, swarming pheromones, alarm pheromones, aggregation pheromones, territorial-marking pheromones and spacing pheromones etc. (For details See : Awasthi, 2013).

THE REPRODUCTIVE SYSTEM

The reproductive systems consist of paired gonads *viz.* the testes of the male and the ovaries of the female, paired genital ducts into which the sexual products are discharged and a median duct *viz.* the ejaculatory duct in the male and the vagina in the female. In conjunction with genital ducts are found specialized glands and organs for storage of spermatozoa. The gonads and genital ducts are of mesodermal origin and the ejaculatory ducts and the vagina are of ectodermal origin, lined with cuticle. In most of the insects, the sexes are separate and the sexual dimorphism is distinct i.e. the male and female individuals can be easily identified by examining the tips of the abdomen.

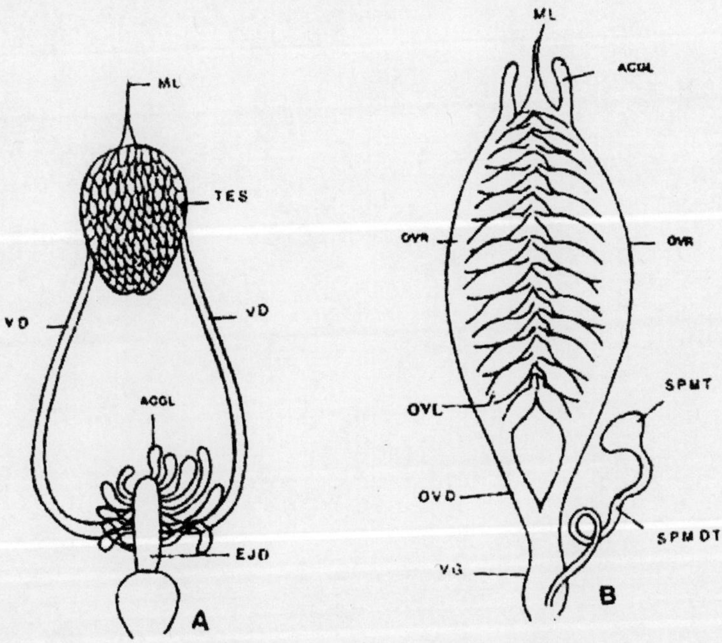

Fig. 21. A. Male reproductive system of a Grasshopper, **B.** Female reproductive system of a Grasshopper *Abbreviations* - ML - medial ligament, TES - testis, VD - vas deferens, ACGL - accessory gland, EJD - ejaculatory duct, OVR - ovary, OVL - ovariole, SPMT - spermatheca, OVD - oviduct, VG - vagina, SPMDT - spermathecal duct.

The Male Reproductive Organs

The male reproductive organs (Fig. 21A) consist of a pair of testes, a pair of sperm ducts *viz.* the vasa deferentia and the ductus ejaculatorious. In most insects there are also a number of accessory glands which open into the vasa deferentia or ejaculatory duct.

The Female Reproductive Organs

The female reproductive organs (Fig. 21B) consist of a pair of ovaries, a pair of lateral oviducts and the median oviduct formed by the union of two lateral oviducts. The median oviduct opens posteriorly through a slit like opening the gonopore. Besides these, there is usually a saccular spermatheca for reception and storage of spermatozoa and a pair of accessory glands of diverse function and a copulatory pouch of the body wall. The latter is either an open genital chamber or a tubular exit passage from the median oviduct termed as vagina.

3

THE DEVELOPMENT AND METAMORPHOSIS

THE DEVELOPMENT

The development of the embryo takes place after oviposition but in some cases it has started and partly taken place in the parental body. In the case of viviparous insects, the whole development takes place within the insect body. The period of the egg stage after oviposition varies to a great extent e.g. it is only momentary as a larva emerges immediately in some cases. In others, it varies from a few hours to a few days, but in some extreme cases it is about 9 months and also it may last for two years. The development is studied in two parts (1) embryonic development and (2) post embryonic development. The embryonic development occurs inside the egg in which development of various organs – systems take place. After the completion of the embryonic development, young ones of the insect emerges. Insects follow different post embryonic developmental patterns. In the simplest, the newly hatched insect is like a miniature adult. It grows and moults till it reaches adult size. In incomplete metamorphosis an immature nymph hatches, grows, develop wings and by stages becomes an adult. Complete metamorphosis involves (1) egg, (2) larva, (3) pupa or resting stage and (4) adult. During development from the egg to adult, the changes in form are known as metamorphosis (see below). For details of embryonic development (see: Awasthi, 2009).

THE METAMORPHOSIS

Insects like other arthropods attain their maximum size by undergoing a succession of moults or ecdyses. The number of moults through which an insect passes is fairly constant for a species. The form assumed by the insect between any two ecdyses is termed as instar. Thus the life history of insects is made up of succession of instars culminating in the adult form. In many of the simple and generalized insects, the several instars are very similar to one another and only differ from their appropriate adults in the absence of wings and the incomplete development of their reproductive system. However, in the majority of the winged insects the adult is in sharp contrast to the wingless young stage. It is these forms that are said to undergo metamorphosis. It is thus clear that the term metamorphosis relates to the development of these insects in which the young stage is completely different in form, size and structure from the adult.

During larval development, there is usually no marked change in body form, each successive instar being essentially similar to the one preceding it, but the degree of change from last instar to the adult varies considerably and may be very marked. This change is called metamorphosis (Snodgrass, 1952, Wigglesworth, 1965). In physiological terms, it is a change which accompanies a moult in the absence of juvenile hormone. According to Highnam and Hill (1979) the overall change in form between the first larval instar and the adult is called metamorphosis.

Types of Metamorphosis

Most insects undergo changes in shape or form during their development, a process of metamorphosis. Some species have only minor modifications, but a gradation may be seen upto very drastic alterations e.g. from legless worm like immature individuals to the adults with legs and wings as seen in flies. The metamorphosis has been categorized in different ways.

Here the term (1) ametabola (2) hemimetabola and (3) holometabola have been used to illustrate metamorphosis.

1. Ametabolous insects

In most primitive insect groups such as Collembola, Thysanura and other Apterygota, no distinct external changes are evident except in size i.e. the youngs are practically identical with adults except in size and reproductive ability. So little change occurs in them during their growth that they are considered as having no metamorphosis. Hence, this entire group is called Ametabola. Examples of ametabolous insects are Spring tail, Silver fish etc. (Fig. 22A)

2. Hemimetabolous insects

In hemimetabolous insects the metamorphosis is incomplete or gradual, the immature instars are usually similar to the adults in general appearance of body, mouth parts, antennae, compound eyes and legs and also in feeding habits. The immature forms differ from the adults in lacking wings and well developed reproductive organs (Fig. 22B).

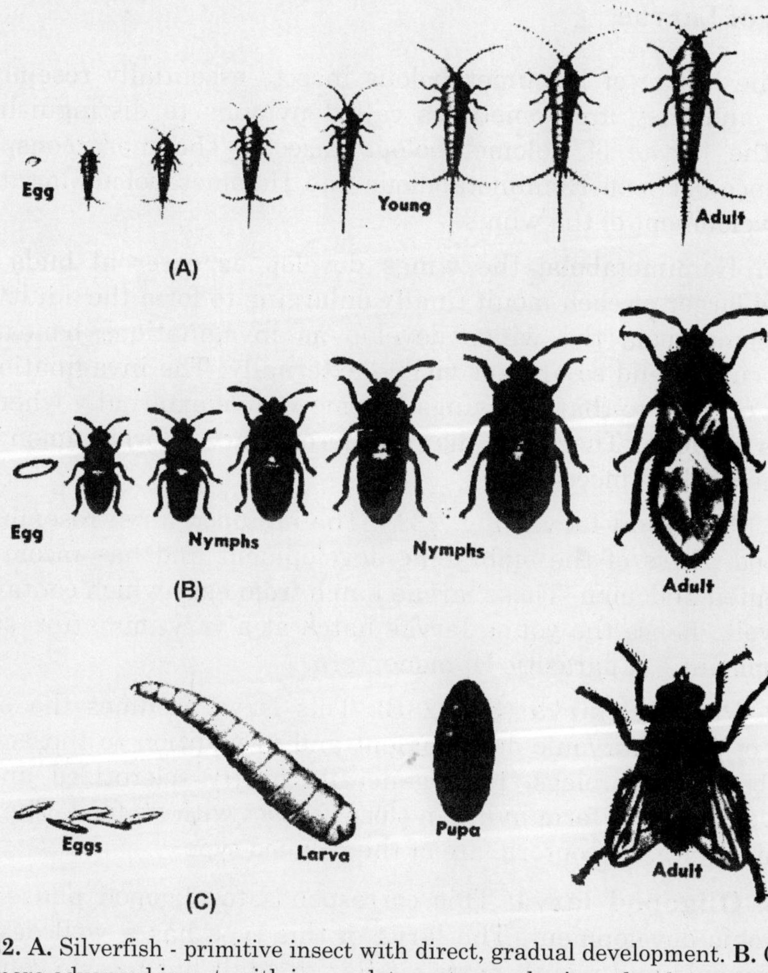

Fig. 22. A. Silverfish - primitive insect with direct, gradual development. **B.** Chinch bug - more advanced insect, with incomplete metamorphosis or development in three stages, **C.** House fly - advanced, modern insect, with complete metamorphosis or development in four growth stages.

3. Holometabolous insects

In holometabolous type of metamorphosis, the wings are developed internally until the preadult instar, in which stage the wings are everted as large pads. The holometabolous metamorphosis is represented by three distinct growth forms during the post embryonic period *viz.* (1) the larva or a feeding stage, (2) the pupa or resting stage and (3) the adult or reproductive stage. The adults have fully developed wings and reproductive organs. Examples of Holometabolous insects belong to Lepidoptera, Diptera, Coleoptera, Hymenoptera, etc. (Fig. 22C).

Types of Larvae

The larvae of Hemimetabolous insects essentially resemble the adults and they are sometimes called nymphs to distinguish them from the larvae of holometabolous insects. The most conspicuous difference between Hemimetabolous and Holometabolous larvae is in the development of the wings.

In Hemimetabola, the wings develop as external buds which become larger at each moult finally enlarging to form the adult wings. In Holometabola the wings develop as invaginations beneath the larval cuticle and so are not visible externally. The invaginations are finally everted so that the wings become visible externally when larva moults to pupa. There are many different larval forms amongst the Holometabolous insects viz.

1. Protopod larva (Fig. 23A): The protopod larva resembles the protopod phase of the embryonic development and has incompletely segmented abdomen. These larvae hatch from eggs which contain very little yolk, hence the young larvae hatch at a very immature stage of development e.g. parasitic Hymenoptera.

2. Polypod larva (Fig. 23B): This larva assumes the polypod phase of the embryonic development and in addition to thoracic legs, has abdominal prolegs. It is generally poorly sclerotized and is a relatively inactive form living in close contact with its food. The larvae of Lepidoptera, Mecoptera, are of the polypod type.

3. Oligopod larva: This corresponds to oligopod phase of the embryonic development. The larva of this type has a well developed head capsule and mouth parts similar to adult but has no compound eyes. They are of two subtypes *viz.*

(a) Compodeiform larva: The larva is well sclerotized, dorsoventrally flattened and is usually a long-legged predator with a prognathous head. The compodeiform larvae occur in the Neuroptera, Trichoptera, Strepsiptera and some Coleoptera (Fig. 23C).

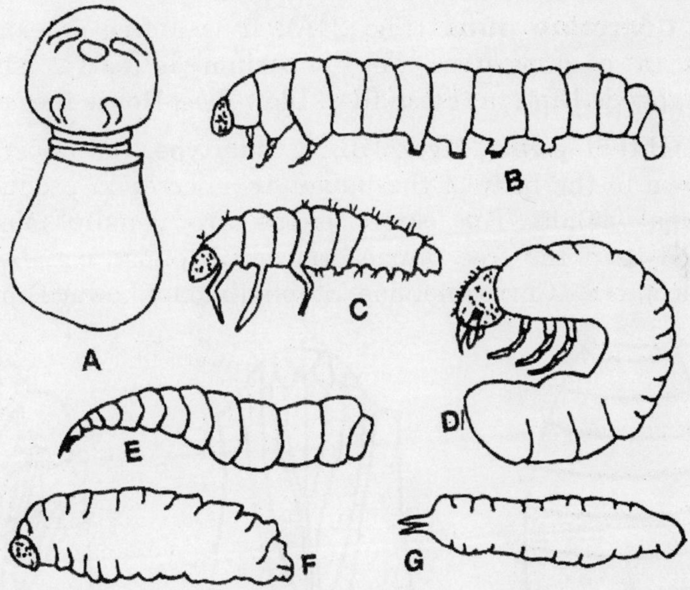

Fig. 23. Different types of larvae. *Abbreviations* - A - Protopod, B. Polypod, C. Campodeiform, D. Scarabaeiform, E . Acephalous, F. Eucephalous, G. Hemicephalous.

(b) Scarabaeiform larva: The larva is fat or robust with a poorly sclerotized thorax and abdomen and which is usually short-legged has got burrowing habit in wood or soil. The scarabaeiform larvae are found in the Scarabaeoidea and some other Coleoptera (Fig. 23D).

4. Apodous larva: As the name indicates, the larva has no legs and is very poorly sclerotized. They have been classified into 3-subtypes on the basis of sclerotization of the head capsule viz.

(1) Acephalous: Without a head capsule, characteristic of Cyclorrhapha (Diptera) (Fig. 23E).

(2) Hemicephalous : With a reduced head capsule which can be retracted within the thorax e.g. Tipulidae and Brachycera (Fig. 23G).

(3) Eucephalous: With a well sclerotized head capsule as found in Nematocera, Buprestidae, Cerambycidae and Aculeata (Fig. 23F).

Types of Pupae

The pupa is a resting or quiescent stage, a characteristic feature of all holometabolous insects. Pupae of some of the insects are quite active e.g. mosquitoes.

Three types of pupae have been recognized.

(1) Coarctate pupa (Fig. 24A): It is a free or exarate pupa enclosed in a puparium. The puparium is barrel shaped e.g. Cyclorrhaphous Diptera (Flesh flies, Blow flies, House flies etc.).

(2) Obtect pupa (Fig. 24B): In this type, the appendages are glued down to the body of the pupae by a secretion produced at the larval/pupal moult. The obtect pupae are usually more heavily sclerotized than the free pupae or exarate pupae e.g. Lepidoptera, some Coleoptera, (Chrysomelidae, Staphylinidae) Lower Diptera.

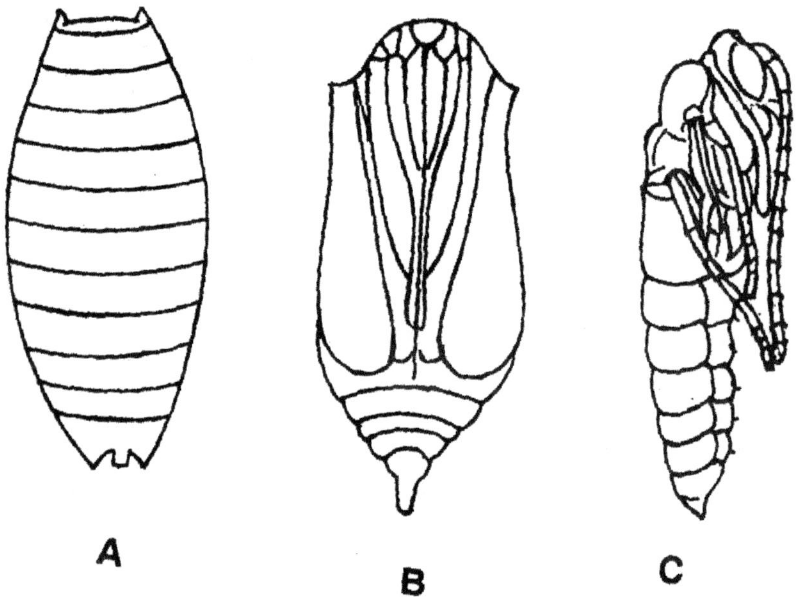

Fig. 24. Different typs of pupae. *Abbreviations* - A. Coarctate, B. Obtect, C. Exarate.

(3) Exarate pupa (Fig. 24C): In this type, the appendages i.e. legs, antennae and wing-pads lie free and are not glued to the body e.g. Hymenoptera, Coleoptera, Neuroptera and Trichoptera.

Further, a number of insects retain parts or nearly complete portions of the previous moulted skin, this phenomenon is called the pharate condition. The pupae in general exhibit only limited movement and most are sessile, found in protected environment such as in the ground, under bark, in cases or retreats, fashioned by the larvae or in silken cocoons produced by last instar larvae.

4

CULTURAL CONTROL

By cultural control is meant those methods of planting, growing and harvesting a crop which will prevent or lessen insect damage. This method consists of a slight departure from the usual time of planting, sowing or harvesting the crops or changing the plan of crop rotation, weed control, disposal of the crop remnants after harvest. Planting good seeds and resistant varieties are very important in controlling some insect pests.

Since cultural methods are usually economical, they are specially useful against pests of low unit-value crops. Practices which reduce the chances of buildup of pest populations may hold them below the level which will cause economic damage. Such methods are particularly applicable to field crops and forests.

Knowledge of the life history or bionomics of a pest species is essential to the effective use of cultural control methods. The principle of the "weakest link" or most vulnerable part of the life cycle usually applies. The environment is changed by altering farming practices at the correct time so as to kill the pests or to slow down their multiplication. In this way, the method is aimed more at prevention than at cure. If the environment is unfavourable, the pest may not reach a population level which will cause serious damage.

Cultural controls are often used when chemical or biological methods have not yet been devised for an injurious species. Cleanup of the sources of infestation and changes in the planting or harvesting time are particularly important when no effective method of killing the pest is known. However, these methods are also used in combination with other controls.

1. Control by planting pest resistant varieties: Seeds of healthy plants preferably of a resistant variety only should be sown as far as possible. Certain varieties are naturally resistant or less susceptible to insect attack e.g. certain varieties of wild apples are immune to aphid attacks while most of the cultivated varieties are very susceptible. Hence, by interbreeding wild with cultivated variety a strain can be evolved which will yield good fruits and shall also be immune to pests. Similarly the Indian cotton *Gossypium orboreum* is much more resistant to the attack of jassids, whiteflies and the pink bollworms than the American cotton, *G. hirsutum*. Also the indigenous cotton is resistant to leaf roller but susceptible to pink boll worm and the foreign variety behaves in an opposite manners. Hence, by crossing the Indian species with the American species a new strain, immune to aphids, jassids and pink bollworms etc. can be developed.

2. Tolerance to insect infestation : A variety may be infected by insects yet survive and show less injury than others because of its ability to replace injured parts such as leaves and rootlets.

3. Ploughing in relation to insect control : The plough if used at right time is a good tool for combating many insect pests; it disturbs or kills them, eradicates weeds upon which they might feed and breed, exposes them to natural enemies or to weather control and buries them so deeply that few adults can emerge. A thorough stirring of the soil before planting is an indirect method for controlling the corn root aphids, because, it breaks up ant colonies, kills many aphids and prevents the growth of weeds on which they live until corn roots are available.

4. Clean cultivation : Clean cultivation means the removal of weeds, plant residues, and other materials from the fields and growing only healthy crops. The destruction of crop residues is a very good preventive measure for controlling certain species of insects e.g. caterpillars or beetles. The elimination of remaining effectively stops further insect breeding by cutting of their food supply and shelter.

Many insects infesting crops will develop on weeds which may or may not be related to these crops botanically. Hence, weed control is a good practice for reducing insect infestation. Damage due to *Leptocorisa* is very serious when paddy fields are surrounded by weeds which provide a good shelter to the pests till the ears have been formed in them.

Similarly various species of red hairy caterpillar, *Amsacta moorei*, *Amsacta collaris* and other species of this insect are polyphagous and lay eggs on a number of weeds and on emergence the caterpillars feed on the weeds and then migrate to the nearby fields of any crop.

5. Crop rotation : Crop rotation is effective as a preventive measures against insects that feed on relatively few plant species or where the insects are incapable of long distance migration but crop rotation is useless against a general feeder. If one and the same crop is grown every year the insect population is bound to increase due to abundance of food material. On the contrary, if crops are grown in rotation or alternate years e.g. a crop of one plant family followed by that of different family e.g. barley grown in spring may be followed by legumes or pulses and then wheat in winter and so on. The crops of the same family e.g. cotton, and lady's finger (okra) belonging to the same family Malvaceae must not follow same year. Mixed crops are sometimes useful as pests do not thrive in large number.

6. Flooding or irrigation : Where water is available, it is sometime possible to destroy insects by flood or irrigation of the field. Flooding rice fields destroys many sugarcane borers. Similarly other burrowing insects e.g. crickets, grasshoppers, beetles and bugs come out of the ground after flooding the fields. They either die their natural death or readily picked up by birds. Ploughing followed by heavy irrigation within 30 days destroys many pink boll worms. The sugarcane and wheat crops can be saved from the attack of termites by irrigation.

7. Fertilizing : Nitrogenous fertilizers in some soils tend to increase the susceptibility of sorghum to damage by attack of some insects. As a result of proper manuring the plant growth is stimulated. The healthy and vigorous plants can easily resist the attack to which the weaklings would have surrendered.

8. Pruning and thinning : Some pests are normally carried from old crop to the new one. Pruning and thinning decrease the intensity of attack. Infested plant shoots and dead hearts must be immediately removed. Ratooning should be avoided and there should be plenty of interval between harvesting a crop and sowing a new one.

9. Time of planting and harvesting : Time of planting and harvesting has great influence on insect infestation e.g. early maturing cotton is not attacked by pink boll worm in areas where moth do not emerge early. Late planted wheat escapes oviposition by hessian flies in America because the plants do not emerge until most of the flies have disappeared as they are very short lived. Similarly early sowing of rice in the Punjab between the 3rd week of May and mid June is helpful in protecting it from the attack of rice borer, *Tryporyza incertulas* so the time should be so regulated that the crop may not be worth the attack at the active period of the pests.

10. Closed season : An insect species limited in its feeding to one crop may be eradicated by not growing this crop for a year or two.

This has been done with success in isolated areas for eradicating the pink boll worm. Such a closed season consists in not growing cotton for a period of several months.

11. Destruction of volunteer crops : Volunteer crops are those which grow from self sown seeds, and stubble crops are those which sprout from roots or stubbles (stumps), may be a source of insect infestation and they should be destroyed by ploughing or other means.

12. Trap crop : A trap crop is a small planting often only a few rows, made somewhat earlier than the main planting for the purpose of diverting insects away from the main crop. The trap crop can either be harvested early or cut and used as fodder before a generation of the insect can be completed. Otherwise it will serve as a breeding ground for the pest which will then attack the main crop in large numbers. The plant species used for trap crop should be one that is very attractive to the insect. The preferred host plants can also be grown around the main crop and when the pest has appeared it can be cut and destroyed.

The other methods of cultural control are early cutting of crop, rate and depth of planting, sowing, separation of complimentary host plants, pasturing the heavy infected crops and drying out the soil etc.

5

PHYSICAL AND MECHANICAL CONTROL

The protection of crops and foods from insect attack by physical and mechanical means is the simplest and most effective method. Such measures consist of destruction of insects by mechanical means, burning, trapping, protective screens or barriers, use of high or low temperature, soaking in water and drying.

1. Hand picking : When only a few plants are infected, certain large conspicuous species may be removed from the plants by hand and destroyed just by pressing the abdomen or dipping them in kerosenized water.

2. Mechanical means: A rotary blow has been found to destroy over 96% of white grubs as well as cut worms, army worms and other caterpillars. The cutting wheel is thirty two inches in diameter and fourteen inches wide and carries sixteen steel blades. It operates by power take off from a tractor at 150 rpm.

3. Trapping : The insects may be turned into a tray by light and baits or the trap may be mobile and drawn across a field.

4. Light traps : The attraction of nocturnal insects to light is a common observation and the light traps have been invented and tested for killing crop pests. The attractiveness of an artificial light depends upon its candle power and colour. A yellow light will attract 60% fewer insects than a white light of the same candle power and brightness. The insects attracted by light may be destroyed as they hit electric grid or they may be drowned in water and oil after they have been turned into the trap.

5. Bait traps : Baits are a form of control whereby insects are attracted to a selected spot on which they can be easily removed. Bait traps have been found to be very effective in controlling the nocturnal insects like cockroaches, crickets, caterpillars and nocturnal moths.

6. Temperature treatments: (a) Low temperature control: Low temperature extremes are fatal to insects of stored grains. Usually a temperature of 28°C or lower for at least several hours will kill most of the insects. *Tribolium confusum* can be killed in 24 h when exposed to subzero temperatures. Low temperatures are utilized for the control of insects in mills, warehouses and similar establishments.

(b) High temperature control: The *Tribolium confusum* cannot complete its development at 41°C and temperature of 44°C or above kills the adults. High temperatures are also fatal to other species and can be used to kill them. Heating cotton seeds to a temperature of 65.5°C for 30 seconds will kill all pink boll worms in the seeds. Temperature of 48.8-51.6°C in all plants of a flour mill for 10-12 h will kill all insects exposed to such temperature.

Exposing the infested grains to the sun in summer also kill all the adults of the stored grain insects.

7. Physical barriers : The protection of crops from insects by barrier is possible in case of crawling and migratory insects. Treated paper barriers and dust barriers are also used. Trenches are also used to stop army worms. Similarly water channels can be used for checking the migration of crawling insects.

8. Burning : Although destroying insects by burning their shelters is sometime practicable, burning pastures and woodland to destroy insects should not be recommended. It has been found that such practices reduce soil fertility and increases soil erosion. Furthermore, it fails to destroy injurious species but does eradicate many beneficial forms.

Both cutting away dead hearts and burning them in time save valuable crops from destruction. Burning the field refuse and stubbles, are the other useful ways of pest control.

9. Sieving and winnowing : It is possible only for domestic purposes.

10. Flooding : It is well known that a large number of insects breed and dwell in the soil, which may be destroyed by flooding the fields with water. There are several other simple methods through which a number of pest infestation can be controlled but the mechanical method of insect pest control involves a lot of labour like shaking the trees and killing the insects either in the larval or adult stages.

6

BIOLOGICAL CONTROL

The term biological control means control of insects by several biological agencies such as parasites, predators and pathogens. The introduction of sterile males in a normal population and use of sex attractants have been reported to bring about effective biological control. De Bach (1964, 1974), Reay (1969), Huffaker (1974), etc. have given exhaustive account on Biological control; a brief account from these books is given here:

The theoretical basis of the biological control depends on the existence of a natural enemy. The effective natural enemy can be deduced to have the following characteristics: (i) high searching ability, (ii) high degree of host specificity or preference, (iii) good reproductive capacity relative to the host and (iv) good adaptation to a wide range of environmental conditions. The most essential characteristic is high searching ability. It should be borne in mind that a really effective enemy may be scarce in its native home because it regulates the host population at low level. There usually is one best enemy for each species in a given habitat and one frequently is sufficient for complete biological control, often, however, a second or third enemy species may add to host population regulation and may in fact be necessary to achieve satisfactory biological control. The best enemy species may differ for different host habitats. Hence there is generally no single best natural enemy extending throughout the range of a pest species. With these qualities the introduced natural enemy by itself or in conjunction with other mortality factors is expected to prevent outbreak of the pest species or atleast to drop down major population fluctuations.

The essential principle of biological control is to maintain a state of biological equilibrium of a living organism. This biological equilibrium is normally maintained by a set of forces called environmental resistance.

Often the biological equilibrium of a living organism is disturbed either due to environmental or biotic factors and consequently a particular organism multiplies and spreads in a unlimited manner so as to assume the status of a pest. The inherent ability of an organism to survive and to reproduce within a given time and under optimal environmental conditions is known as biotic potential. Biotic potential is a force opposed to environmental resistance. It takes into account the number of young produced per female in each reproduction, the number of reproduction in a given time and the sex ratio of the species.

Advantages of biological control

Only a limited number of introduced pests have been successfully exterminated with chemicals. It would seem wiser in many instances to attempt a biological control programme which might reduce pest populations below the level which causes economic damage. Biological control has the tremendous advantage that, if successful, it becomes self-sustaining and intergrated into the normal environment of the control area.

If pest populations are reduced to the point where economic damage is negligible, the control programme is a success. Biological controls tend to be particularly useful on low unit-value crops where complete control may not be required or where chemicals are not recommended. Pests of field and forage crops, forests, and range may be economically controlled by biological methods; chemical controls against such pests may be impractical.

A biological control destined to be fully effective will be easily and quickly established. If an imported parasite or predator is not established within three years of careful releases under good conditions for its development, the programme may be justifiably discontinued.

TYPES OF BIOLOGICAL CONTROL

There are 4 main types of biological control viz. (i) introduction of parasites, (ii) predators, (iii) pathogenic organisms such as bacteria, viruses and fungi etc. and (iv) removal of fertile males and creating sterility in males by gamma radiation and introducing these sterile males in a normal population. In addition to these 4 main types, there is a fifth type of biological control where the indigenous parasite of the locality is being utilized for this purpose.

(1) Parasites (Fig. 25)

Parasitic insects develop as larvae on or in a single host individual from eggs generally laid on, in or near the host and usually consume all or most of the host body, killing the host and then pupate, either within or outside the host. The free living adult parasite emerges from the pupa and starts the next generation a new by actively searching for host in which to oviposit. They tend to attack only one host stage i.e. eggs, larvae, or pupae, although there is also some overlapping in certain cases, adult insects do not serve as hosts. Very often life cycles are commonly short, ranging from 10 days to 2 weeks or so in mid summer but correspondingly longer in cold weather. In general, they all have great potential rates of increase.

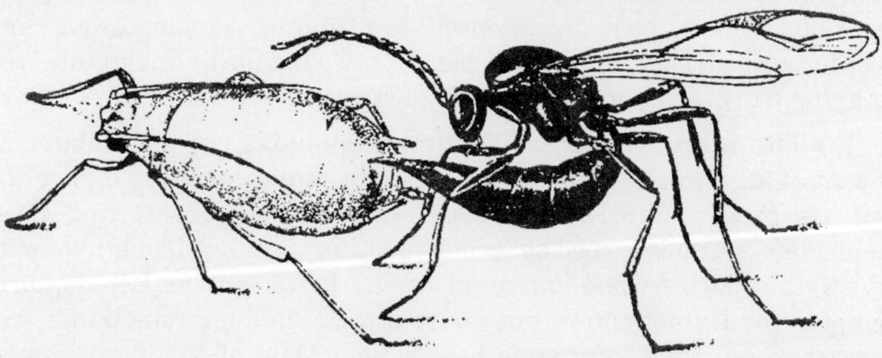

Fig. 25. A minute parasitic wasp, *Aphidius testaceipes,* laying egg in an aphid (From Pfadt, 1962)

According to the species one or more parasitic larvae may characteristically develop per host individual, thus we have either solitary or gregarious parasitism. Sometimes two different species of larvae develop from eggs laid in the same host. This is multiple parasitism. There is also ectoparasitism and endoparasitism, depending upon whether the larva develops externally or internally. The major groups occur in two orders of insects viz. Hymenoptera and Diptera.

HYMENOPTERA

This is the dominant order among all entomophagous insects. Over two thirds of the cases of successful biological control of pests species have been achieved by hymenopterous parasites. The most important and numerous hymenopterous parasites occur in the major groups commonly known as chalcids, braconids, ichneumonids and proctotrupeds.

(1) The chalcids : The chalcids attack species in nearly all orders of insects, the preferred ones being Coleoptera, Diptera, Homoptera and Lepidoptera which also happen to include the bulk of our chief crop pests. The immature stages of chalcids predominantly develops in host's eggs or larvae or pupae and some develop in adults of Homoptera. The encyrtids and aphelinids include many that specialize in attacking scale insects, mealy bugs, white flies, and related forms and have been so outstandingly successful in biological control.

(2) The braconids: This is a major large family of parasitic wasps. They have been used in biological control. Only an occasional rare species is hyperparasitic hence the family is nearly entirely beneficial. Parasitism may be external or internal, solitary or gregarious. Their host preferences are mainly lepidopterous and coleopterous larvae but also commonly Diptera, including the damaging fruit flies and Homoptera, particularly aphids.

(3) The ichneumonids: The Ichneumonidae comprise about 20 per cent of all parasitic insects. Most are primary parasites, hence the family is predominantly beneficial. Some of the largest and most conspicuous parasite species occur in this family, Ichneumonidae probably parasitize wood and stem boring larvae in the Lepidoptera, Coleoptera or Hymenoptera but many attack lepidopterous pupae too. Some species have a very wide host range. Most of the Tryphoninae are solitary parasites of sawflies. The Ophioninae are predominantly internal parasites of lepidopterous larvae although a few develops in the grubs of root feeding scarabs in the soil.

(4) The serphoids : This contains several families, the most important of which are the Platygasteridae and the Scelionidae. All are parasitic on immature stages of other insects and they are predominantly primary internal parasites.

DIPTERA

In addition to tachinids which comprise the most important dipterous family in biological control, this order contains several parasitic families including - Cyrtidae, Nemestrinidae, Pipunculidae and Conopidae etc. However, tachinids constitute the major parasitic family both in numbers of species and in economic importance and have been the only group utilized extensively in biological control.

The Florida red scale, *Chrysomphalus ficus* invaded Israel and spread throughout the Citrus growing areas and soon become a major citrus pest. An indigenous ectoparasite, *Aphytis chrysomphali* was the only parasite commonly recorded from the scale in Israel. The attacks of this species were limited mainly to male scale pupae and its effect on the scale population was negligible.

Another species of *Aphytis* viz. *Aphytis holoxanthus* and the endoparasitic *Pteroptrix smithi* were introduced into Israel from Hong Kong during 1956-57 against the Florida red scale; and complete biological control was achieved throughout the coastal region, which is the main citrus growing area of Israel, during the first 2 or 3 years following the introduction of the parasite, and the Florida red scale was virtually eliminated from that area as a pest.

Similarly biological control of Olive scale was achieved by introducing two additional species of parasites viz. *Coccophagoides utilis* and *Anthemus inconspicuus* from Pakistan to California. The biological control of California red scale was achieved by introduction of *Aphytis lingnanensis* from South China to California.

A great deal of success has been achieved in Newzealand and other parts of the Australia by using *Pteromalus puparum* imported from United States to control populations of the small white butterfly, *Pieris rapae*.

Although various Aphidius species have been used for aphid control, the best known Braconid, is *Macrocentrus ancylivorus*. It has been reared on a large scale in the United States using the larvae of the Strawberry leaf roller as the host and liberated for the control of Oriental fruit moth, *Grapholitha molesta* in peach orchards.

Finally, mention should be made here of the spectacular eradication of the Citrus black fly, *Aleurocanthus woglumi* from Cuba and various islands in West Indies by *Eretmocerus serius* imported from Malaya. Control was achieved so rapidly that when *Aleurocanthus woglumi* become an important pest in Mexico some years later, *Eretmocerus serius* was first choice as control agent. Unfortunately, it was a dismal failure and further investigations indicated that climatic conditions were unfavourable. The pest was eventually brought-under control by *Amitus hesperidum* and *Prospaltella* species imported from India.

FACTORS GOVERNING PARASITE

One of the important factors governing parasite introduction for biological control is the geographical location of the area under experimentation and the conditions met within that area and presence of an indigenous fauna of a peculiar and restricted character. Introduced parasites meet with relatively little competition from indigenous forms. Limitation of the area to be covered by parasitic colonization, resulting in the presence of only few main crops and therefore facilitating adequate organization and centralization of control experiments. Other factors which hamper an efficient programme of biological control are the phenomena of multi- parasitism and hyperparasitism, there is ample evidence to show that the efficacy of introduced parasites has been enormously reduced by these two factors.

Certain other considerations which have a direct bearing on parasitic introduction are suitable adaptation of artificial measures of control so as to put the destruction of more parasites than hosts, proper recognition of the time factor in biological control experiments and proper evaluation of the capacity of a parasite to discover its host.

Utilization of indigenous parasites

Biological control is also brought about through the agencies of indigenous parasites of a locality, such operations consist either of conserving or increasing the number of parasites in a given area or of attempting to colonise such parasites by artificial culture methods under laboratory experimental conditions in an area of a country where they did not previously exist.

One of the important features of this type of biological control is the alteration of the host parasite ratio in a given locality. Example of such a biological control is *Trichogramma minutum*, a chalcid egg parasite on notable pests such as *Pyrausta nubilalis* and the sugarcane borer, *Diatraea saccharalis*.

(2) PREDATORS (Figs. 26-30)

Predator insects differ from parasitic ones in that the larvae or nymphs, as the case may be, require several to many prey individuals to attain maturity. The adults generally deposit their eggs near the prey population and after hatching the active mobile immatures search out and consume prey individuals. This need of larvae or nymphs to search is an important distinction from parasitic forms whose larvae develop on a single host individual and thus have the advantage of not having to discover additional hosts. Adults of many species are also predatory. There may be one or several generations to one of the prey. Both larvae and nymphs as well as corresponding adults may be predaceous or only one stage may exhibit the habit.

1. Coleoptera: The main predaceous families of the Coleoptera are - Coccinellidae, Staphylinidae, Histeridae, Lampyridae, Cleridae,

Cantharidae, Meloidae, Cicindelidae, Carabidae, Dytiscidae and Gyrinidae. The Coccinellidae and the Carabidae are of most important in biological control of agricultural pests.

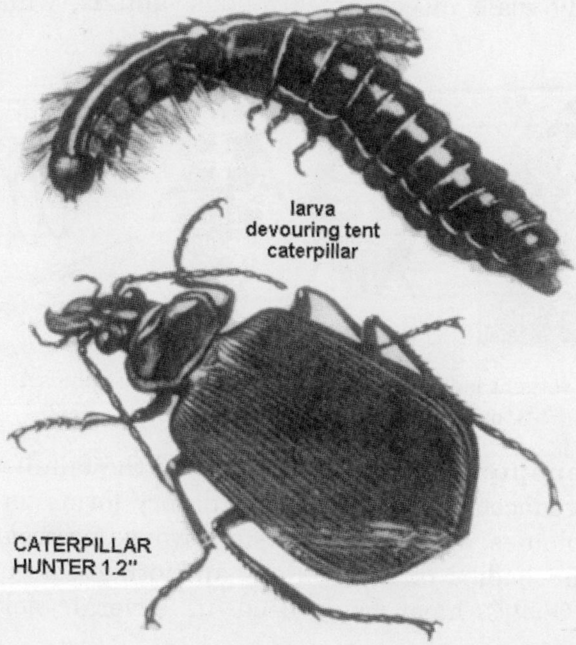

Fig. 26. Caterpillar hunter: Beetles and larvae all are predators feeding on insects and other small animals.

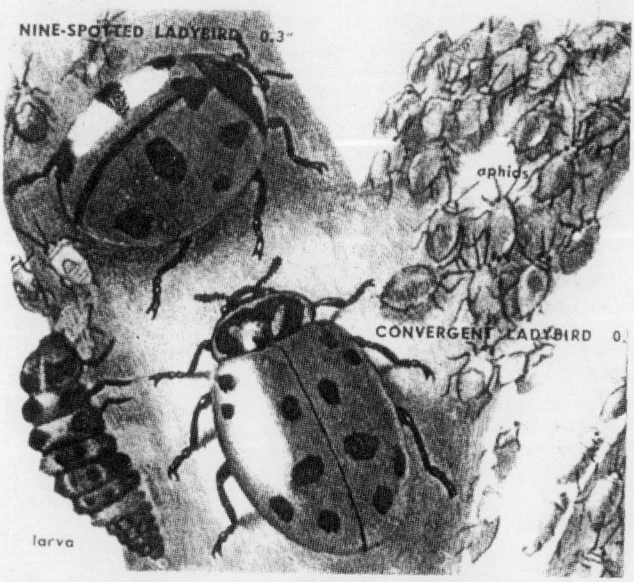

Fig. 27. Adults and larvae of Lady bird beetles feeding on aphids.

2. Neuroptera: Most of the species of the order Neuroptera are predatory. The most important families in biological control have been the Chrysopidae and Hemerobiidae which attack many agricultural pests including scale insects, mealy bugs, aphids, white flies, mites etc.

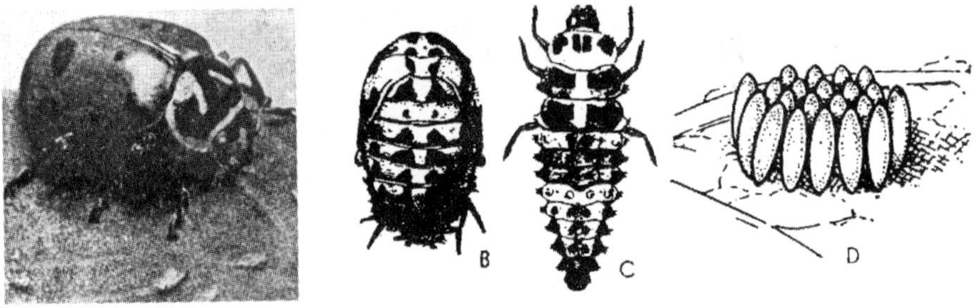

Fig. 28. Convergent lady beetle, a common predator of aphids, A. adult beetle; B. pupa, C. larva, D. eggs. (From Pfadt, 1962).

3. Hymenoptera: About one quarter of the families of this order are strictly predaceous. Most of the predatory forms tend to be social and live in colonies. The colonial vespid, wasps provision their paper nests with the bodies of caterpillars and other soft bodied insects. Species of *Polistes* have been used in several biological control projects.

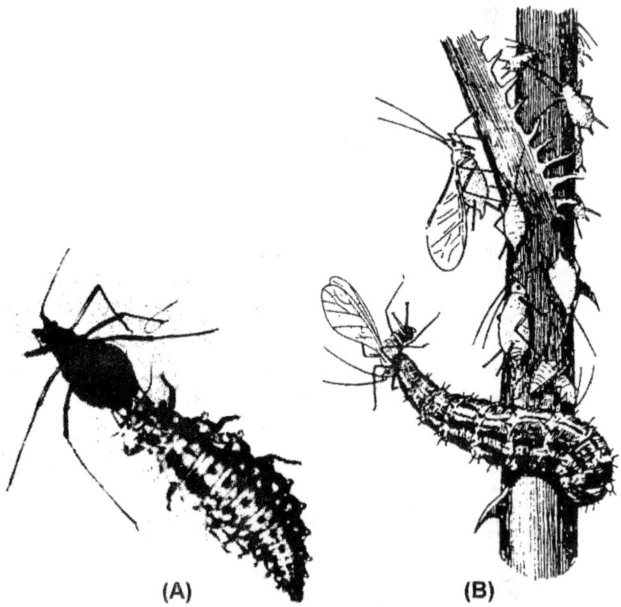

Fig. 29. Two efficient predators of aphids. **A.** lacewing larva feeding on aphids, **B.** syrphid maggot feeding on aphids (From Pfadt, 1962)

4. Diptera : A few families are entirely predaceous and many others contain predatory species. The more common or important economic families include the Syrphidae, Asilidae, Cecidomyiidae, Anthomyiidae, Calliphoridae and Sarcophagidae. Of the latter four, many are predaceous in the egg masses of grasshoppers.

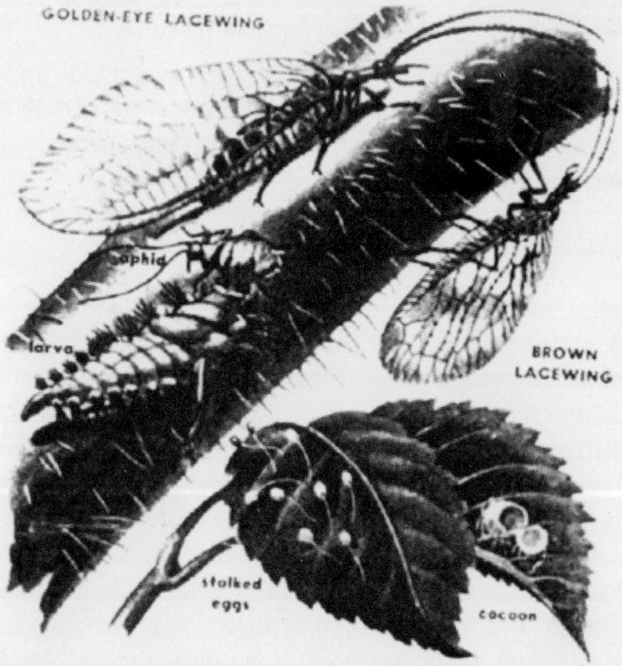

Fig. 30. Lacewing larva feeding on aphid

5. Odonata : The dragon flies and damselflies are predaceous both as nymphs and as adults. The nymphs are all aquatic and feed on a wide range of aquatic insects.

Biological control has been very successful in certain cases where injurious insects have become established in land, they did not previously inhabit and have developed into persistent pests. A classical examples of such a biological control is that of the cottony cushion scale, *Icerya purchasi* in California, by a predatory coccinellid beetle, *Rodolia cardinalis* introduced from Australia. Introduction of this beetle in California resulted in almost complete and permanent destruction of *Icerya*, within less than 5 years. This species was so completely destroyed that the survival become numerically insignificant. Subsequently the predator achieved complete biological control of the cottony cushion scale in 25 additional countries and a substantial degree of control in four other countries (De Bach, 1964, 1974).

Many outstanding successes in the biological control of serious coccid pests have since been recorded. Some notable cases include the complete biological control of the Citrophilus mealy bug, *Pseudococcus fragilis* in California, the coconut scale - *Aspidiotus destructor* in the island of Fiji, Maritius and Principe, Green's mealy bug, *Pseudococcus citriculus* in Israel; the red wax scale, *Ceroplastes rubens* in Japan; and Coffee mealy bug, *Planococcus kenyae* in Kenya.

A mention may be made of the vertebrate predators like fishes, frogs, reptiles, insectivorous birds and mammals etc. which are the regular feeders and deserve to be provided with suitable conditions to grow and breed profusely.

(3) PATHOGENS

Pathogenic microorganisms attack insects and have life cycles more or less characteristic or similar microorganisms developing in other groups of animals. Insects are probably subject to as wide as variety of disease as are the vertebrates. Except for the fungi, disease organisms gain entry in the host via mouth or the digestive tract, i.e. the insect host must eat plant or other food contaminated with pathogen. In case of fungi, entrance is gained through the insect integument and free water or very high humidity is generally required. Thus, fungi tend to be restricted to moist environments. However, fungi do have advantage of attacking sucking insects which because of the nature of their feeding on sap, tend to be fairly free of disease caused by microorganisms because they rarely ingest them. Virtually no insect disease organisms occur in mammals and none have been recorded from man. Thus they are safe to use in biological control even in large scale microbial spraying operations. The most common disease of insects are caused by bacteria, fungi, viruses, and protozoa.

1. Bacteria : Because it can be produced fairly easily on a large scale and has a wide spectrum of activity, *Bacillus thuringiensis* was of the first microorganisms to be incorporated into a commercial "microbial insecticide". The most sensitive insects appear to be lepidopterous caterpillars, particularly those in which alkaline conditions prevail in the midgut. This is related to the fact that the spores (it is a crystal bearing spore former and can be conveniently stored and applied in the resistant state without loss of potency) can only enter the insect on ingestion and that the proteinaceous crystals of toxin is only liberated at around pH 9-10.

Those pathogenic bacteria which do not form spores have proved to be of little practical value in the field and indeed only two other groups of spore formers have been employed with any success i.e. *Bacillus popilliae* and *Bacillus cereus.* The former species produces a condition known as milky disease in the Japanese beetle, *Popillia japonica* and has been used extensively in North America for control of this pest and Chaffer beetles. Although *Bacillus cereus* does not produce a crystalline toxin, it also requires closely defined mid gut conditions with an optimum around pH 7. This *Bacillus* is active against the codling moth and certain Hymenopterous pests.

2. Viruses: The classic example of the employment of viruses in pest control is the highly successful use of a polyhedrosis viruses against the pine sawfly, *Neodiprion sertifer* in Canada. A related pest *Neodiprion swainei* and the spruce sawfly, *Diprion hercyniae* are similarly controlled by a polyhedrosis virus.

Another example is the control of alfalfa caterpillar, *Colias eurytheme* by introduction of polyhedrosis virus. The cabbage looper, *Trichoplusia ni* and other brassica caterpillars have been controlled in a similar way. Only one other group of viruses, the granulosis type has been used against pests in field, in this case the Lepidopterans, *Pieris brassicae* and *Pieris rapae.* Further, for *Nucleopolyhedroviruses,* which are used for the control of insect pests (see: Harper, 2012).

3. Fungi and Protozoa

Pathogenic fungi have several properties which should make them ideal organisms for use in biological control. They produce highly resistant stages and are capable of prolific spore production once they became established. However, their great dependence on microclimatic conditions, especially humidity has seriously curtailed their use and only a few have been employed successfully of these – *Entomophthora exitialis* has produced promising results against the spotted alfalfa aphid, *Therioaphis maculata.*

Many protozoa appear to cause chronic rather than acute symptoms in their victims and the microsporidian, *Thelohania hyphantriae* takes upto four weeks to exert its full effect which has been shown to cause mortality in the fall webworm, *Hyphantria cunea.*

4. GENETIC METHODS

Genetic pest control implies the manipulation and use of genetic material in a manner injurious to pest insects. The control of insect pest populations by the release of sterile males has been demonstrated with at least five insect species (Lachance *et al.,* 1967). This spectacular technique has been termed autocidal control and involves using an insect species to bring about its own self-destruction. It is accomplished by irradiating laboratory reared males of the species to an extent sufficient to disrupt the genetic function of the sperm nucleus but not appreciably interfere with the normal ability of the male to mate or of the sperm to penetrate the egg of the female. However, such fertilized eggs fail to develop so a wild female mated with a sterilized male produces no progeny. If sterile males are released in large enough number in relation to the wild population they will fertilize more females than will the wild males. The advantage is cumulative in each generation, hence eradication may be achieved within a few generations under ideal conditions. The major success thus far has been with the screw worm which was eradicated from the I'sland of Curacao, from Florida, and from Texas to California by massive annual release of sterile males.

A low infestation of the Oriental fruit fly was eradicated from Guam in 1963; the melon fly was eradicated from Rota Island in 1963 using a combination of a bait spray and sterile insect release, and the field cockchafer eradicated from a seventy four acre plot in Switzerland in 1962. The boll weevil is reported to have been eradicated from a small isolated field by the release of chemically sterilized males. Chemosterilants have been under intensive investigation but are not as yet practical. They offer the advantage of treating the wild population directly and thus avoiding the mass rearing necessary with gamma radiation (De Bach, 1964, 1974).

USE OF PHEROMONES

The ability of an insect to locate a mate and to copulate once it has been found, is often directed by natural specific chemicals produced by the insect called pheromones. Natural chemical attractants and mating stimulants have been shown to occur in many pest insects. They are often effective in incredibly small concentrations; one caged virgin female of the introduced pine sawfly attracted over 11000 males. Some have been chemically identified and even synthesized. Its high specificity for particular target insects however makes this approach highly desirable. Such phenomena gave rise to the intriguing idea of using these natural or artificial pheromones to attract and trap insects, to lure them to contact poisons, chemosterilants or pathogens and to mask the location of females by saturating the environment with synthetic sex pheromones. Such techniques could offer highly specific control methods with little or no ecological side effects. An example of possibility is furnished by the eradication of the Oriental fruit fly on Rota I'sland where the artificial pheromone or attractant viz. methyl eugenol lured the male flies to a poisoned coated surface. A similar but more important achievement was the eradication of the mediterr-anean fruit fly from about one million acres in Florida in 1956-57. It appears that adequate control from pheromones or attractants probably can only result if they are so highly effective as to be able to reduce the pest population to low level in the treated area and are used over a large area so that immigration of the pest from the outside would not negate the effort. Pheromone traps are commonly used for controlling various moths *Heliothis, Spodoptera, Leucinodes* etc.

7

INSECT HORMONES AS PESTICIDES

It is well known that the growth and moulting in insects is controlled by three main hormones. The brain hormone secreted by the neurosecretory cells of the brain, activates the prothoracic glands. These glands in turn secrete one or several closely related steroids, the ecdysones. The ecdysone(s) cause the insect to moult. The kind of cuticle secreted by the epidermal cells at each moult is affected by a third group of hormones, the juvenile hormone, which are secreted by the corpora allata. In the presence of juvenile hormone, larvae – moults into larvae. However, in their absence in hemimetabolous insects larvae metamorphose into adults and in holometabolous insects, larvae moult into pupae and then into adults.

Moulting is controlled by regulating the release of brain hormone. Maturation and metamorphosis are controlled by regulating the release of juvenile hormone. In simplest terms, the ecdysone or its analogues stimulate the synthetic activities necessary for moulting, whereas juvenile hormone analogues influence the kinds of synthetic activities that occur in response to the ecdysone. Out of the three the last two hormones and their analogues or mimicking substances are of much value as for as the control of insect pest population is concerned.

A. Use of Juvenile Hormone as insecticides

As is known, the juvenile hormone inhibits metamorphosis in all insects and as long as juvenile hormone is present in high concentration larvae will moult into larvae and they may die in this way without attaining adulthood.

It was the discovery of the Slama's "paper factor" (Slama and Williams, 1965, 1966) that opened a new chapter on the methods of the control of insect pests, by inhibiting metamorphosis and causing several other abnormalities. The greatest attention has been paid to juvenile hormone analogues with reference to their possible utilization as entirely new types of insecticides. The juvenilising effect of juvenile hormone analogue is very great. The application of 0.001 mg of the juvenile hormone analogues to the body surface of the last instar nymph of *Pyrrhocoris apterus* or very brief contact with the "paper" leads to a complete juvenilising effect. It has been also pointed out that the "paper factor" named 'juvabione" extracted from the North American trees such as the balsam fir (*Abies balsamea*) normally used for manufacturing the paper has been extremely useful in controlling the *Pyrrhocoris apterus* and other members of the family Pyrrhocoridae. It has been clearly demonstrated that the application of 0.01 mg of the crude extract of juvabione, a juvenile hormone analogue extracted from balsam fir, is quite effective and inhibited metamorphosis of the last instar nymphs of *Pyrrhocoris apterus*.

The delayed action of juvenile hormone on metamorphosis after application to the late embryo was also seen in *Pyrrhocoris*. Normally during the moult to the fifth instar the corpus allatum stops secreting juvenile hormone and the larva moults to an adult 7 days later. If however, juvenile hormone analogue is applied at the outset of fifth instar, then the individual moults after 5 days to a supernumerary sixth instar larva. Similarly appliction of juvenile hormone analogue(s) to the *Pyrrhocoris* embryo after blastokinesis also produced sixth instar larvae. Some of these giant larvae moulted again to seventh instar larvae but they never metamorphosed (Riddiford, 1972).

It has also been calculated that about 1 gram of the crude extract of the balsam fir tree would be sufficient to kill about 100 million bugs. Likewise, it has been found that topical application of 0.1 mg juvabione to *Pyrrhocoris apterus* sterilizes the adult females. Slama and Williams (1966) reported complete inability of certain insect species to reproduce in contact with a juvenile hormone analogue or "paper factor" present in the American paper products. After the discovery of the "paper factor" there occurred a rapid progress in the research of juvenile hormone analogues. As remarked by Slama *et. al.* (1974) the general principles concerned with practical use of juvenile hormone analogues, in the insect control occur to be quite simple. The basic idea is that the development disturbances caused by juvenile hormone analogues at different ontogenetic stages may lead to mortality of the affected species/or at least may prevent insects from reproduction. Since a real potency of insect pest gradiations is primarily associated with their tremendous capacity to reproduce, it is easy to realize as pointed out by Slama *et. al.* (1974), that now serious consequences for development of insect pest population such an arrest of reproduction might have with juvenile hormone analgues. The applied Entomologists get an access to chemical growth regulators which have no direct toxicity to insect's and other form of life i.e. human beings, domestic animals, birds etc. Thus juvenile hormone analogues form a new class of chemicals which can be characterised as specific regulators of insect growth and development. Williams (1967) suggested the term "third generation pesticides" to the methods of the hormonal control of insect pest population particularly the use of juvenile hormone(s) as insecticides and a realistic account of the practical use of juvenile hormone analogues in insect control has been expressed by Staal (1971).

Further, juvenile hormone analogues not only inhibit metamorphosis in adults but also block the maturation of imaginal discs which are the primordia of many adult integumentary structures in holometabolous insects. The imaginal discs occur in larvae as tiny sacs of epithelial cells which come directly from embryo e.g. in Lepidoptera, each imaginal wing disc is set aside in the embryo as a small nest of cells destined to form the pupal and adult wing. These imaginal disc cells divide during larval life but fail to secrete a cuticle until the time of pupation.

Pupation is brought about by a falling concentration of juvenile hormone and results in the rapid morphogenesis of wing disc and the secretion of a cuticle. If juvenile hormone is applied to a Lepidopteran larva prior to pupation, it prevents both the morphogenesis and cuticle synthesis in the wing disc. When juvenile hormone disappears, the imaginal wing disc continues its development (Slama and Williams, 1966). Riddiford and Williams (1967) and Riddiford (1970) have demonstrated that the application of juvenile hormone analogues to young embryos interferes with their development. Thus juvenile hormone blocks not only the metamorphosis of the larva to adult but it also blocks the equally profound metamorphosis of the embryo to larva. Thus it is clear that juvenile hormone analogues play a significant role in the embryonic development.

An interesting feature of juvenile hormone action is that they act directly on the target cells and do not require mediation of any specialized organ. An other feature is that the juvenile hormone analogues rapidly penetrate the cuticle of insects.

It has been found that the immature insects which have been affected by an exogenous juvenile hormone usually fail to moult spontaneously, die soon after moulting or fail to reproduce (Schneiderman, 1972). Juvenile hormone analogue treated insects are intermediates between larva and pupa, pupa and adult, or larva and adult.

Bhatnagar-Thomas (1972) has studied the potentiality of juvenile hormone analogue as pesticide in controlling *Trogoderma granarium*. She has reported the inhibition of metamorphosis and has recommended the use of juvenile hormone analogue for control of Khapra beetle.

Jacob and Prabhu (1979) have reported various types of abnormalities in the embryo of *Dysdercus singulatus*. Awasthi (1985) has reported that juvenile hormone analogue treated last instar larvae of *Philosamia ricini* indicated numerous abnormalities e.g. 5-7 days duration of larval life was prolonged to 24 days and they died without metamorphosis. The pupae retained larval characters and the head, wings, antennae etc. were abnormal and adult if emerged were also abnormal and died soon after emergence.

Schneiderman (1972) has observed that young or immature insect larvae are not killed by juvenile hormone analogues. In order for these agents to kill insects they must act during last larval instar or during metamorphosis where they cause the production of intermediates. Alternatively, juvenile hormone analogues must act upon adults where they function as chemosteriliant. Furthermore, juvenile hormone analogues must be applied at the proper time and must persist for several days to be effective in producing non-viable intermediates.

As pointed out by Schneiderman (1972) this matter has been analysed by many workers. The main argument is that : to produce maximum kill with minimum dose, juvenile hormone analogues must be applied when some parts of the insects body are sensitive to these substances whereas other parts are not e.g. it has been found that bugs, locusts and other exopterygote insects usually become insensitive to juvenile hormone analogues a few days after the last larval moult. Consequently one must apply juvenile hormone analogues shortly after the last larval moult, generally during the first third of the last larval instar or much later, after adult emergence, where juvenile hormone analogues will have an ovicidal effect (Schneiderman, 1972). Riddiford (1972) has also studied the ovicidal action of juvenile hormone analogues and has remarked that the juvenile hormone analogues can act as an ovicide if given to adult female or to her freshly laid eggs and this action of juvenile hormone would seem to be the most practical means of using juvenile hormone for insect control. One can eliminate the potentially harmful larval stages as well as next generation and even if the application is not early enough to prevent hatching the delayed effects will prevent metamorphosis into a normal reproductive adult.

B. Use of moulting hormones as insecticides

A brief mention may be made here also of the use of moulting hormone in pest control. Studies of several workers have shown that most insects die soon after they have been induced to moult by an injection or topical application of an ecdysone. Alpha ecdysone appears to be less toxic than the other ecdysones, perhaps because the insect regulates its rate of conversion to the active form, beta ecdysone. The most striking abnormalities produced by ecdysones are juvenile hormones like effects e.g. if a pupa of *Tenebrio* is injected with 25 micrograms of beta-ecdysone it moults into a second pupa instead of into an adult. It has been also found that exogenous ecdysones usually kill insects raises the question of whether ecdysones may be suitable as insecticides. They affect insects at all stages from newly hatched larvae to adults. They can be applied topically in various solvents and penetrate the cuticle and several volatile derivatives have been synthesized so that ecdysones can be used as fumigants. Some analogues are effective when fed to insects. Moreover, many ecdysones are quite stable. Also various conjugated ecdysones are being synthesized in an effort to produce ecdysone which the insect cannot break down. Robbins *et al.* (1968) have demonstrated that various conventional insecticide synergists like piperonyl butoxide synergize the action of ecdysone analogues.

Another attractive feature of the potential use of ecdysones and related substances for insect control is that they are already widespread in the environment. It was discovered that substances with ecdysone activity are found in many vascular plants in some cases accounting for as much as 1% of the dry weight of certain plant tissues. Thus for more than 40 different steroids with ecdysone activity in insects have been isolated from higher plants by Nakanishi *et al.* (1966), Takemoto *et al.*, 1967), Siddall (1970) etc. These phytoecdysones include authentic alpha and beta ecdysone.

Thus, it is hoped that in future synthetic hormone analogues both juvenile hormone analogues and moulting hormone analogues would be utilized for controlling the pest populations after solving some of the practical difficulties.

8

PLANT QUARANTINE AND LEGAL CONTROL

Legal control is the lawful regulation of areas to eradicate, prevent or control infestation or reduce the damage caused by insects. This mainly involves the use of quarantines and pest control procedures. The central and state officials often work with local community or districts as in mosquito or locust control projects.

No attention was paid to the insect pests being carried from one place to another with the transportation of plants or seeds till it was noticed that the insects introduced into the foreign countries thrive best and increased at a very high rate in their new homes e.g. when American grapes were brought to France, *Phylloxera* caused such a heavy damage to French vine yards that the necessity of checking this pest from being carried to various parts of the world was seriously felt. Hence legislation had to form laws for this purpose. India and western countries like USA have passed a number of Acts to regulate the pest and to transport plant and their products. Under these Acts all imported goods should be carefully inspected to be insect free or should be fumigated and sterilized before allowing them to disembark. The export and import laws of the country now enforce the compulsory fumigation of food stuffs before and after embarkation.

The purpose of plant quarantine is to prevent by legal restrictions the entrance and establishment of a plant disease or insect pest in an area where the disease or pest does not exist. The area to be protected by this means may be a country, a state or it may be a smaller unit such as a part of the state. There are four methods of plant quarantine viz. inspection at the point of destination, inspection and certification at the point of origin, complete embargoes and controlled introduction of plants.

1. Inspection at the point of destination: This inspection is necessary because many insect pests can be discovered and their accidental introduction thereby eliminated. Often it is physically not possible to examine all materials likely to carry insect pests with sufficient care to detect them. Also because some insects and plant diseases have inconspicuous stages they are extremely difficult to locate.

2. Inspection at the point of origin : Most countries and many states follow this plan i.e. inspection and certification at the point of origin. In this case materials are allowed to enter the state or the country as the case may be, provided they bear a certificate issued by the plant quarantine officer which mentions that the materials have been inspected and found to be free from insect infestation or plant disease infection. Again here also it is not possible to seperate physically all the materials likely to carry pests.

3. Embargoes : By this we mean the exclusion of all plant materials or commodities which are classed as hosts of the insects. This is also not possible in practice. Because the area where the pest exists cannot always be definitely determined and the material listed in embargo may not include all of those which might carry or harbour the pests.

4. Controlled introduction : This consists of allowing the importation of only a very limited amount of material for propagating purposes, obtaining it from pest free or disease free areas and from plants believed to be insect or disease free, inspecting the shipment very carefully then planting in an isolated place for a preliminary period. After obtaining healthy stock in this way, it can be released for multiplication and distribution.

Enforcement of plant quarantine

1. Foreign plant quarantine : Plant quarantine inspectors are stationed by the Government at all of the principal ports of entry. Materials which must be watched are those which can carry the pests or diseases. They not only include the food plants of the insects but any article on which any stage of it might be found. Such as cotton used for packing, gift might contain seed infested with the pink boll worm and straw which could be infested. The central and state Governments have enforced several laws for this purpose, from time to time.

The Government of India have enforced several Acts like sea customs Act of 1876 to stop the entry of Mexican cotton boll weevil, destruction Insect Pests Act of 1914, Poison Act of 1919, Drug Act of 1940 etc. for preventing the entry of infested seeds and not allowing any material without proper inspection and certification that the material is free from all shorts of insect infestation as well as fungal, viral and bacterial diseases. As a rule all the food stuffs are fumigated.

9

CHEMICAL CONTROL

Recent discoveries of new synthetic insecticides have sparked exciting advances and major breakthroughs in the control of insect enemies. Chemicals have subdued pests that once caused wide spread crop destruction, death of domestic animals and epidemics of insect borne human diseases. The modern insecticides are both effective and reliable. The whole world is resorting to them more and more for the solution of many insect problems.

The chemicals which kill the insects by their chemical action are termed as insecticides. They are used for the protection of men, domestic animals, crops, agricultural products from the attack of insects when other methods fail to control the pests. Insecticides are seldom used in full strength but are formulated in a variety of ways to make them easier for application.

FORMULATIONS OF INSECTICIDES

Following are the common formulations of insecticides viz., (1) dusts, (2) granular formulations, (3) insecticide-fertilizer mixtures, (4) wettable powders, (5) solutions, (6) emulsifiable concentrates, (7) aerosoles, (8) fumigants and (9) miscellaneous formulations.

Whatever may be the formulations, the poisonous chemical present in an insecticide must penetrate the vital organs and tissues of the insect and ultimately kill it.

(1) Dusts : Insecticidal dusts are those powders which are used dry and mixed with or impregnated with certain organic materials or pulverised minerals (powders) such as talc, pyrophyllite, bentonite etc. These minerals are called carriers or vehicles since they carry the insecticide. Dusts are blown to deposit on plants by dusting machinery or blowers. Ground to a fine size, most dust will pass through a 325 mesh screen and range in size from 1 to 40 μ. The finished dust may be 0.1 to 25% active material. In dust form, in general, the toxicity of an insecticide increases as the particle size decreases.

(2) Granular formulations : These are similar to dusts except for larger particle size. The range of particle size in a granular product is designated by a two figures mesh classification e.g. 30/60 means that virtually all the insecticide granules will pass through a standard 30-mesh sieve while a negligible quantity will pass through a standard 60 mesh sieve. Some of the common granular formulations are 16/30, 20/40, 24/48 and 30/60. Granular insecticides are generally used as dressings on or in the soil and may be applied with fertilizer spreaders or special granule applicators.

(3) Insecticide fertilizer mixtures : Insecticide fertilizer mixtures may be formulated by adding granular insecticides to commercial fertilizers or by spraying insecticides directly on to the fertilizer. Such mixtures are applied at the regular fertilizing time to provide both plant nutrients and control of soil insects.

(4) Wettable powders : Theses are similar to dusts but they are meant to be diluted and suspended in water and used as spray. To make an insecticidal dust act in this manner a dispersing and wetting agent is added to the formulation. They are more concentrated than dust as they may contain as high as 75% toxicant.

(5) Solutions : Many of the modern synthetic insecticides are insoluble in water but soluble in organic solvents. These soluble insecticides in solution form are used directly for insect control. They are however seldom used on plants because of their phytotoxic reaction.

(6) Emulsifiable concentrates : The most common and versatile formulation is the emulsifiable concentrate. This formulation consists of an insecticide, a solvent (for the insecticide) and an emulsifying agent. Mixing the concentrate with water forms an emulsion. The solvent used may evaporate quickly after spraying, leaving a deposit of toxicant after the water has evaporated.

The use of an emulsifying agent serves several purposes; it makes possible the diluting of a water insoluble chemical with water, it reduces the surface tension of the spray thus allowing it to spread and wet the treated surface more effectively, helps the spray and make a better contact with the insect cuticle.

Generally oil soluble emulsifying substance is used. Normally emulsions are unstable and break up into their component parts. This action is termed breaking. For spraying on plants a quick breaking mixture is preferred. Since this results in heavier deposits of toxicant.

(7) Insecticidal aerosols : Aerosols are minute particles suspended in air or fog or smoke. The diameter of these particles range from 0.1 to 50 μ. The dispersion of insecticide into aerosol form may be accomplished by burning, vapourising with heat, automizing mechanically or releasing through a small hole an insecticide that has been dissolved in a liquified gas. The last method of aerosol preparation gives the popular aerosol bomb.

(8) Fumigants : Insecticides used in the gaseous form are known as fumigants. Fumigants are most often formulated as liquids under pressure and are held in tanks. When the liquid is released in open air it changes back to a gas, quite often fumigants are a mixture of two or more gases.

(9) Miscellaneous formulations: There are certain insecticides which are special formulations meant for specific uses. Often insecticides are stuffed in large pills and capsules and introduced into the stomach of animal. Insecticides may be mixed in shampoos, intended for use on house pets, waxes for use on floors may contain an insecticide. Poison baits consist of toxicants combined with a food stuff attractive to the insect pests. All these types of insecticides are special formulations designed for special purposes.

SYNERGISTS

Some chemicals have the property of greatly increasing toxicity of certain insecticides. When the increased toxicity is markedly greater than the sum of the two used separately, it is termed a synergistic action. Most synergists have been used with pyrethrum or allethrin etc. Synergistic action is important because it provides a means for a more effective insecticide and it reduces the cost of control.

CLASSIFICATION OF INSECTICIDES

The foregoing types of insecticide formulations are all lethal, effective in killing the insects sooner or later. Generally the insecticides are broadly classified on the basis of the mode of their entry into the insect body as stomach, contact and fumigant poisons. Modern classification however does not follow this basis since the present day synthetic insecticides enters the insect's body in more than one way. To avoid this overlapping modern classification is now based on the chemical nature of insecticides and the major divisions are therefore inorganic and organic. The organic insecticides are further divided into oils, botanicals and synthetics. Oils are petroleum products used primarily against fruit tree insects. Botanicals are of plant origin and include such insecticides as Rotenon, Pyrethrum etc. Among the synthetic group of organic insecticides, several chlorinated hydrocarbons, organophosphorus compounds etc. used in the past are now banned.

With regards the older classification, stomach poisons are materials which are ingested by the insects and which kill primarily by the action on or absorption from the digestive system. Usually they are limited to the control of chewing insects. Contact poisons are absorbed through the body wall and must come in direct contact with the insects to kill them. They are usually adopted for destroying sucking insects. Fumigant poisons enter the tracheal system of the insects in the form of gas. Insects such as stored grain pests and certain soil insects are killed by the fumigants.

1. INORGANIC INSECTICIDES

Though inorganic insecticides have been largely replaced by organics, some of the forms are still used to control certain insects. Most of the inorganic insecticides are stomach poisons. Inorganic materials used in insect control are chiefly compounds of Arsenic, Fluorine and Sulphur.

Note: Most of the arsenical and fluorine compounds have been banned for domestic and agricultural use in many countries. The main compounds are: 1. Lead arsenate, 2. Calcium arsenate, 3. Paris green or aceto arsenate of copper, 4. White arsenic, 5. Zinc arsenate, 6. Sodium fluoride, 7. Barium fluosilicate and 8. Sodium fluoaluminate.

Sulphur Compounds

Amongst the sulphur compounds finely ground sulphur has been widely applied as a dust for control of mites and other insects. Sulphur is boiled with freshly slaked or hydrated lime to produce a mixture known as liquid-lime sulphur, one of the chief constitutents of this mixture is calcium polysulphide. Liquid lime sulphur is used in the control of scale insects, aphids, etc. on fruit trees.

Another important inorganic insecticide is Bordeaux mixture which is prepared by mixing copper sulphate, lime and water in a definite proportion. This is chiefly used as a insect repellent against cucumber beetles and leaf hoppers.

There are many other inorganic compounds which are employed as insecticides. Sodium salinate applied in the soil acts as a systemic insecticide and is used primarily to control aphids on ornamental plants. Sometimes mercurous chloride (calomal) and thallous sulphate are also used as insecticides for controlling cockroaches and ants.

2. ORGANIC INSECTICIDES

1. Oils

These form one group of organic insecticides which in their natural stage are highly phytotoxic. However, when they are used in an emulsion they may under certain conditions be safely applied to plants. They vary greatly in the degree of refinement and in physical properties according to their intended use. Oils are employed in a number of ways; they may be used as solvents or carriers for insecticides. Oils by themselves are insecticidal and there are different types of oils.

(a) Summer oils are highly refined and less phytotoxic. They are generally used to control mites and scale insects.

(b) Dormant oils are less refined and are also applied to control scale insects.

(c) Superior spray oils are chiefly meant for controlling Aphids.

(d) Emulsible oil containing 95-99% oil and emulsifier.

(e) Tank mix oils are made up of oil, emulsifier and water added separately to a spray tank. The last two are used for the protection of agricultural crops.

The use of oils has several advantages, initially it is cheap, have a good spreading capacity, are easy to mix and quite safe to animals life. Another important consideration is the fact that insects have not developed so far any resistance to oils. However, there are certain disadvantages with oils e.g. they have a low toxicity to insects. They are unstable and extremely phytotoxic.

2. Insecticides of Plant Origin

These are insecticides derived from plants and majority of them are contact poisons. They have several uses in insect control. The most important insecticides of plant origin are nicotine, pyrethrum, rotenone, sabadilla, ryania and neem, all of which serve as insect toxicants.

(1) Nicotine : It comes chiefly from two plants, *Nicotiana tabacum* and *Nicotiana rustica*. They are extracted from the leaves and stems of these plants Nicotine is highly toxic to a great number of insects acting on the ganglia of the central nervous system possibly at the synapses. It is toxic when ingested, absorbed through the body wall or taken in through the trachea. The most commonly used form today is nicotine sulphate at 40% content. Free nicotine solution is also used as a fumigant or greenhouse spray. Nicotine insecticides are used as 0.04% sprays 3% dust, fumigants and dips. Water or summer oil may be used as a diluent in sprays and hydrated lime, ash or sulphur for preparing dust. Nicotine sprays and dusts are safe to use on plants. They are compatible with most other insecticides.

(2) Pyrethrum : This derives its toxicity from 4-esters: pyrethrin 1, pyrethrin 2, sinerin 1 and sinerin 2. It is a plant product extracted from the flowers of *Chrysanthemum cinerariaefolium.* A synthetic compound recently developed known as allethrin has the same toxicity as pyrethrum.

It is prepared by extracting the toxins with kerosene, alcohol or ethylene dichloride from flowers of the *Chrysanthemum.* This concentrated form may be diluted with petroleum oil and used as sprays. Concentrated extracts as acetone, alcohol or a hydrocarbon solvent are available for spraying plants. These are diluted with water to form aqueous emulsion. Pyrethrum is harmless to warm blooded animals and man but is highly toxic to most insects. It is used for controlling the houseflies and mosquitoes. It acts rapidly as a nerve poison and the affected insects fall quickly on to the ground. This is termed 'knock down' and the insects thus knocked down need not necessarily die, some may recover after a lapse of time. Often the toxicity of pyrethrum is enhanced by the addition of chemicals such as sulphoxide or piperonyl butoxide to the insecticide. Such chemicals which increases the toxicity of an insecticide are known as "Synergists".

(3) Rotenone : This is prepared by grinding the dried roots and bark of several species *viz. Derris, Lonchocarpus* and *Tephrosia.* Rotenone may be extracted with a solvent like chloroform or carbon tetrachloride and a concentrated extract is then prepared. This may be diluted in sprays or mixed with a dust carrier and used as an impregnated dusts. It is nontoxic to warm blooded animals and man. It is also non-phytotoxic. But it is poisonous to cold blooded animals. Alkaline carriers such as lime should not be used with it. Rotenone sprays and dusts loose their toxicity rapidly on exposure to light and air. It is a selective insecticide and can kill only some insects i.e. rotenone is very specific in its toxic action to insects. It is believed to kill the insects by inhibiting the utilization of oxygen by body cells or depriving the tissues of oxygen through a depressing reaction on the nerve and muscles connected with tracheal breathing. They are used in combating cattle grubs, lice, aphids and beetles.

(4) Sabadilla : It is derived from the seeds of *Schoenocaulon officinale.* It is sold as a finely ground powder diluted with an inert filler. A 50% concentrate is manufactured. This should be diluted and applied as a dust at 10 to 20% sabadilla content. Sabadilla dust is strongly irritating to the nasal memberanes causing sneezing and eye irritation. These effects are however transitory and soon disappear after exposure. Sabadilla is most useful as contact dust to control Hemipterous insects which are resistant to other insecticides. It is very specific in its action and is ineffective against many other common insects.

(5) Ryania : It is obtained from the stems and roots of a plant called *Ryania speciosa.* This is used for the control of sugarcane borers and certain Lepidopterous insects. It acts both as a contact insecticide, as well as a stomach poison. It has residual effect and is relatively stable.

(6) Neem (*Azadirachta indica* A. Juss): A lot of work has been done in the past few decades and several seminars and symposia were held to evaluate the insecticidal properties of various parts of the neem tree viz. leaves, flowers, fruit kernels in the powder or oil form to control insect pests infesting various crops and crop-products. It has been stated that the Neem is an important source of insecticide and all its parts viz., leaves, bark, flower, kernel and kernel oil are variously used as insect-repellents. Since ancient times, its leaves are kept amongst woollen and silken clothes and books to protect them against insects. Farmers throughout the country are aware of the need to protect the grains from spoilage during storage and use of neem leaves in grain heaps.

The oil is also used as a disinfectant and emulsifying agent in insecticides. The maximum insecticidal activity has been found in neem kernel. Several workers have tried the neem leaves and kernel powder as grain protectant. Mixing of seeds with neem powder at 1.0 and 2.0% effectively protected from pest infestation. Neem oil at 0.5 and 1.0% provided the best surface protection when applied to pulses against the pulse beetles *Callosobruchus chinensis* and *C. maculatus.* Repellent effect of extracts of neem leaves, flowers and kernels against rust red flour beetle *Tribolium castaneum* and Khapra beetle *Trogoderma granarium* has been evaluated and found that neem kernel extract was more repellent than the leaf or the flower. Adults of *T. castaneum* failed to reproduce when fed with flour treated with neem kernel extract. Neem oil was found effective in preventing adult emergence and prolonging the development from egg to adult stage of *Rhizopertha dominica.*

Chopped leaves of neem were found effective in controlling the *Sitotroga cerealella* in stored rice. In addition to stored grain pests, numerous other pests of agricultural importance have been controlled by neem products in the form of powder, seed cake or aqueous and solvent extracts.

Several commercial formulations have been evaluated against pest of tobacco, groundnut, black pepper, pulses, cotton, oil seeds, vegetable and other agricultural crops. Neem plant parts, also act as growth regulator, repellent and antifeedant, thus help to control crop pest population without altering the quality and nutritive value (For details see : Narwal *et al.,* 1997, and Koul and Wahab 2004). One of the most potent derivatives of Neem is Azadirachtin.

Powdered flowers and fruit kernel powder treated rice, caused mortality of *Sitotroga* adults, 59-64% after 7 days, 2-8% powder was used. As pointed out above neem oil 1-2% was found most effective in reducing adult population of *Sitophilus, Rhizopertha* and *Ephestia.* The neem kernel powder when mixed with cowpea *Vigna sinensis* at 3%, caused 85-95% mortality of *Callosobruchus* (For details see : Singh *et al.,* 1997).

3. Synthetic Insecticides

(A) Chlorinated Hydrocarbons : Some commonly used chlorinated hydrocarbon insecticides are described as follows :

(1) Dichloro-diphenyl trichloroethane (DDT)* : It is a very powerful insecticide, which exercises a prolonged lethal residual effect till covered by the dust or dechlorinated. It is a creamy white, crystalline powder having a faint unpleasant odour. It is waxy to touch, quite stable, resistant to oxidation and light. It is almost insoluble in water but soluble in organic solvents like castor oil, cotton seed oil, xylene, benzene, kerosene, dioxane, ether, acetone, chloroform, carbon tetrachloride, liquid paraffin etc. It is not a repellent hence, insects do not avoid it but it acts as stomach as well as contact poison. The effect is gradual but sure and persists for a long time. It is non-phytotoxic except to cucurbits. It is very effective against most insects such as beetles, weevils, caterpillars, thrips,

* Now banned.

whiteflies, and leaf hoppers. For aphids, it is used in emulsion form. For a quicker action, the DDT is mixed with pyrethrum or other chemicals like BHC or malathion in sprays and smokes. It is compatible with most insecticides and fungicides, except lime sulphur. Considering the effective insecticidal dose, it is comparatively less toxic to mammals (a human fatal dose taken orally is 30 gms). Insolubility in water renders it harmless to use on human skin in dust or aqueous emulsion but when dissolved in oils, its large doses may cause a number of diseases like convolution, cerebral depression, and necrosis of liver. Its inhalation is also harmful but due to its high toxicity for insects, large areas can be treated with small quantities of insecticides by ordinary hands or power sprayers. It is never used in the pure form but compounds are made with certain bases to form dust or spray. The dose of DDT ranges between 0.5 to 20% in mineral oils, 0.1 to 5% of emulsified dispersible or wettable powder in water and 2 to 10% in dust mixed with talc, clay, chalk, sulphur, and other carriers. Most plants, except cucurbits do not suffer by treatment with DDT. The DDT analogue, DDD (dichlorodiphenyl dichloroethane) is less toxic to mammals and is used for flies and vegetable pests. The acute oral LD50 for male rats is 113 mg/kg. DDT is stored in the body fat of birds and mammals and excreted in the milk of mammals.

(2) Benzene hexachloride (BHC) Gammexane* : It is a crystalline tasteless, perfectly stable powder of sand colour, having a camphor like odour, practically insoluble in water but soluble in oils and paraffins. It is available as a pure product or mixed with a carrier like lime, gypsum or chalk. It is perfectly compatible with sulphur and DDT but it decomposes, when mixed with alkalies like lime, Bordeaux mixture and nicotine etc. It is five times toxic than DDT hence acts more quickly but being volatile, the residual effect is less lasting. It can serve both as stomach and contact poisons for most of the insects such as thrips, leafhoppers, aphids, woodlice, and earwigs but less effective against bollworms, citrus fly and is more or less ineffective against flour moths. It can be used as a slow acting fumigant for stored products alone or in combination with carbon disulphide, carbon tetrachloride etc. It is less toxic to mammals when taken orally or absorbed to skin. A fatal oral dose for man is 30 gms (its gamma isomer is more toxic and 15 gms is fatal dose for man).

(3) Lindane : Lindane is the coined name for the gamma isomer of BHC of a purity of not less than 99%. Using lindane in formulation,

* Now banned

eliminates the disagreeable odour and tastes on some crops which result when they are treated with BHC. It exhibits strong stomach poison action, persistent contact toxicity and fumigant action, against a wide range of insects. It is non-phytotoxic at insecticidal concentrations. The technical BHC causes "tainting" of many crops but there is less risk of this with lindane. It is stable to air, light, heat and carbon dioxide and is not attacked by strong acids but can be dehydrochlorinated by alkalies. It occurs as colourless crystals and is practically insoluble in water, slightly soluble in petroleum oils, soluble in acetone, aromatic and chlorinated solvents. The acute oral LD50 for rats is 88 mg/kg. It is effective against many soil insects e.g. beetle larvae and adults, fly-larvae, ants and many biting and sucking insects e.g. aphids, psyllides, whiteflies, capsids, midges, thrips etc. This insecticide is now restricted in use. Trade names: Lindane, Lintox etc.

(4) Aldrin*: It is a white crystalline solid, insoluble in water but is quite soluble in mineral oils and highly soluble in most organic solvents like acetone, benzene, xylene etc. It is stable to heat, alkaline and mild acids and can be mixed with alkaline soils, fertilizers and other insecticides like sulphur, Bordeaux mixture, calcium arsenate, nicotine, paris green, rotenone etc. It is broad spectrum, persistent, nonphytotoxic insecticide. It acts as a stomach as well as a contact poison. Toxic effect on insects is cumulative and therefore, it is not advised to use on edible crops. It is used against all soil insects like termites, beetles (both larvae and adults) fly larvae, cutworms, wireworms, crickets etc. Aldrin is dusted on soil at the rate of about 2.5 kg per acre. Aldrin can control a variety of soil insects at the depth of about 10-11 cm. It is used as 30% emulsifiable concentrate, 2.5-5% dust, 5 and 20% granules, 20-25% wettable powder, or liquid seed dressing. The treated seeds should not be used for human or animal consumption.

Insecticidal preparations containing aldrin, should be handled with care. Respirators should be worn by those applying this material and excessive skin contact should be avoided. Acute oral LD50 for rats is 67 mg/kg. it is absorbed through the skin.

(5) Endosulfan : It does not accumulate in milk or fat but has a high mammalian toxicity. In plants and animals both, it is metabolized to a sulphate. It is a brownish crystalline solid, insoluble in water but moderately soluble in organic solvents. Endosulfan is not phytotoxic. It is a contact and stomach insecticide and acaricide. It has

* Now banned

high toxicity to beetles, lepidopterous larvae and mites. It is also effective against sucking insects. It is extremely toxic to fish but is safe for bees. It is formulated as emulsifiable concentrate, wettable powder, dust and granules. Trade names: Thiodan, Endosel etc.

Note: Other products viz. - dieldrin, endrin, chlordane, toxaphene and methoxychlor etc. were used in the past but currently banned.

(B) Organo Phosphorus Compounds : These are extremely poisonous to man and involves special care, precautions, apparatus and protective clothings and rubber gloves for handling them. All systemic insecticides belong to this category. Those still in use are given below:

(1) Malathion: It is a non-poisonous, non-phytotoxic insecticide. The chemical is yellowish brown liquid slightly soluble in water and miscible with most of the organic solvents. It has a strong penetrating, unpleasant smell and is broken up into non-toxic ingradients when mixed with alkaline media. It can be used as stomach and contact poison, besides as a strong fumigant for mites and other insects which are not killed by other poisons. It is non-toxic to man, 100 times less than parathion either orally or by skin absorption or by vapours. It can be used as contact poison for aphids, scale insects, mealy bugs, jassids and caterpillars in 5% dust or 25-50% emulsion for spraying. Two or three applications fortnightly will control white flies, mealy bugs and scale insects. Acute oral LD50 for rats is 2800 mg/kg. Trade names - Malatox, Cythion etc.

(2) Methyl Parathion : It is a white crystalline substance in pure state, sparingly soluble in water and well soluble in most organic solvents. It has a low thermal stability. It hydrolyzes rapidly in alkaline medium. Due to its low stability, toxicity is lost soon from the treated surface. In soil, micro-organisms decompose it within a week. It is a contact and stomach poison with mild fumigant action. It has penetrating action also. Within 3-4 days 80-90% of it decomposes. It can be used against aphids, mites, leafhoppers, whiteflies, scale insects etc. It is available as emulsifiable concentrate and dust. Trade names: Folidol, Dalf etc.

(3) Trichlorphon : It is initially used for controlling house flies, acts as a contact as well as a stomach poison. It is a white crystalline powder, quite soluble in water, insoluble in petroleum oils, poorly soluble in carbon tetrachloride, diethyl ether, benzene and ethanol. It is stable at room temperature but is decomposed by water at higher temperatures and acid media to form dichlorvos. Manufactured in various forms like wettable powder (50%) soluble powders (50 to 80%), granules (2 to 5%), emulsifiable concentrate (50%) and mixed with sugar crystals. Used for controlling ticks, mites, house flies and fly maggots and bugs. Trade names: Dipterex, Tugon etc.

(4) Diazinon: It is a non-systemic insecticide with some acaricidal action. It may be phytotoxic at higher dosages. It is a colourless oil almost insoluble in water but is miscible with ethanol, acetone, xylene and is soluble in petroleum oils. It decomposes above 120°C and is susceptible to oxidation.It is stable in alkaline media but is slowly hydrolyzed by water and dilute acids. It is compatible with most insecticides but should not be compounded with copper fungicides. The acute oral LD50 for male rats is 108 mg/kg. Effective against flies and fly-maggots, mites, thrips, aphids etc. Formulated as aerosol-solutions; 25% emulsifiable concentrate, 40 and 25% wettable powder, 4% dust and 5% granules. Trade name: Basudin.

(5) Dichlorvos : This insecticide has combined actions of contact, stomach and fumigant. On account of its high vapour pressure, it has greater penetration and it gets readily decomposed. When applied as emulsifiable concentrate, it is effective against sucking and leaf-mining insects. It is non-phytotoxic and non-persistent. It is also used as a fumigant against mosquitoes and other dipterous insects. It has low dermal toxicity but it is toxic to bees. It is marketed in the form of an emulsifiable concentrate. Trade names: DDVP, Nuvan etc.

(6) Monocrotophos : It is a crystalline solid with mild ester smell. It is miscible with water, soluble in acetone and ethyl alcohol, but insoluble in kerosine and diesel. It is corrosive to iron, steel and brass.

It is a fast acting stomach, contact and systemic insecticide. It can be used against a wide range of pests. It is absorbed by roots and then translocated to stem and leaves. Scorching of edges occur in some varieties of sorghum, apple and cherry. Monocrotophos is recommended for the control of aphids, white flies, leaf miners, thrips, shoot borers, boll worms, caterpillars, folige feeder-beetles and a number of other pests. It is formulated as emulsifiable concentrate. Trade names: Nuvacron, Corophos etc.

(7) Fenitrothion : It is a brownish yellow liquid with faint odour. It is insoluble in water but soluble in most organic solvents. Fenitrothion is slowly hydrolyzed in the presence of alkali. Its thermal stability is also low and may explode at 100°C. Iron container encourages decomposition. It is a contact and stomach insecticide with penetrating property. It is a low toxic poison. It decomposes in plants quite actively. It is highly effective in controlling sucking pests like aphids, jassids, bugs as well as caterpillars and flies. Used in the form of 5% dustor 50% emulsifiable concentrate. Trade names: Folithion, Fenitox etc.

(C) Carbamate insecticides

Carbaryl : It is a contact insecticide with slight systemic action. It is widely used against the insect pests of fruits, vegetables, cotton and other crops. It does not deteriorate in storage and has a long residual effect on the crops. It is comparatively safe for insect pollinators, parasites and predators. It is formulated as dust and emulsifiable concentrate. Trade names: Sevin, Carbaryl etc.

3. SYSTEMIC INSECTICIDES

These are the insecticides which are absorbed into the sap stream by the plant tissues on application to leaf or root system and are then transmitted to various parts in such quantities as to render its juices sufficiently toxic to certain insects specially sap suckers for several weeks. These are therefore termed as systemic insecticides. Their use in food and fodder crops cannot be recommended but are extremely useful for cotton and ornamental plants. The systemic insecticides can be safely used against phytophagous insects but not against predators and parasites. They may be applied as sprays or granules to either soil or foliage. The concentration is determined with respect to specific pests.

It is important to note that a large number of organophosphorus compounds act as systemic insecticides. These, as a rule, are highly poisonous substances hence full protective clothing must be always worn. Although, majority of them are non-phytotoxic but many are harmful to bees, livestock, fish, game and wild animals. They are mostly used against sap sucking insects and phytophagous mites. Following are some of the common systemic insecticides

(1) Dimethoate

This insecticide is used mainly against fruit flies and aphids. The pure compound is a white solid, only slightly soluble in water but soluble in most organic solvents except saturated hydrocarbons. It is stable in aqueous solution and to sunlight but it is readily hydrolyzed by aqueous alkali. It is incompatible with alkaline pesticides. The acute oral LD50 for rats is 250-265 mg/kg. It is effective against aphids, psyllids, some flies, sawflies, mites etc. It is marketed in the form of 20-40~ emulsifiable concentrate, 20% wettable powder and 5% granules. Trade names: Rogor, Dimetox etc.

(2) Demeton-methyl (Methyl demeton)

It acts as a systemic as well as a contact insecticide and acaricide, with a fast killing ability and moderate persistence. It is a clear yellowish oily liquid insoluble in water and soluble in most organic solvents. The acute oral LD50 for male rats is 65 mg/kg. It is effectively used against aphids, mites on most crops, some flies and has only a limited effect on brassica aphids. It is used as emulsifiable concentrate. Trade name: Metasystox.

(3) Phorate

It is a persistent insecticide used in granular and emulsifiable concentrate formulations for protection of seedlings from sap feeding and soil insects. It is a clear liquid only slightly soluble in water but miscible with carbon tetrachloride, dioxane, xylene, and vegetable oils. It is hydrolyzed by alkalies in the presence of moisture. The acute oral LD50 for male rat is 3.5 mg/kg, for females 1.6 mg/kg. Used in the form of granule. Trade names: Thimet, Foratox etc.

(4) Phosphamidon

This insecticide is readily absorbed by the plants but with only a little contact action. It is non-tainting and non-phytotoxic except to some cherry varieties and sorghum varieties. It is a pale yellow oil miscible with water and is readily soluble in most organic solvents except saturated hydrocarbons. It is stable in neutral and acid media but is hydrolyzed by alkali. Compatible with all but highly alkaline pesticides. It corrodes iron, tinplate and aluminium and is packed in poly ethylene containers. Acute oral LD50 for rats is 28.3 mg/kg. It is effectively used against sap feeding insects and leaf eating ones, particularly aphids, bugs, many caterpillars (but not Noctuidae) rice stem borers, thrips, beetles, grasshoppers, saw flies, fly larvae and mites. Used as 100% emulsifiable concentrate. Trade names: Dimecron, Daron etc.

(5) Quinalphos

It is a white crystalline solid without any odour. It is sparingly soluble in water but readily soluble in most of the organic solvents. It gets readily hydrolized in alkaline medium. It is a strong contact and stomach insecticide and acaricide. It is effective against a wide range of insects, particularly sucking and caterpillar pests. Used in the form of 4% dust, 25% emulsifiable concentrate and 5% granule. Trade names: Quinatox, Desalux etc.

4. FUMIGANTS

These are respiratory poisons taken into the tracheae in gaseous form and exert their influence when absorbed into the tissues from tracheoles. The other fumigants are suffocating agents which simply clog the spiracles and slowly kill the insects by cutting oxygen supply. Such chemicals can be applied to plants in air tight chambers, green houses and nurseries and to plant products and grain houses, mills, granaries etc. Time for fumigating granaries, mills etc. may be prolonged for 6 to 32 hours. It is extremely essential that no one must be allowed to enter the chamber in which operations are being carried out and provision must be made to open several doors, windows and shutters from outside without entering it.

(1) Carbon disulphide

This gas is highly inflammable and explosive when mixed with air, hence its use involves a great risk. Even a lightened cigarette or a spark from an electric switch is enough to ignite it. It is a yellowish or water white, heavy liquid with a disagreeable odour. Because of the fire and explosion hazards, the carbon disulphide should not be used in fumigation, if other fumigants can be obtained. It is very toxic to insects and plants. For seed and grain fumigation carbon disulphide is applied to the top layer of grain, most effectively in the form of a coarse spray from a ground spray pump. As the vapour is heavier than air, it sinks down through the grains killing all form of insect life. The time of exposure is 36 to 60 hours.

(2) Nicotine

It is used for fumigating green houses in humid conditions at dark hours. It can be obtained as such by heating nicotine to liberate the poisonous gas or paper soaked in liquid nicotine and dried may be burnt. Tobacco plants on roasting also emit the same gas. The acute oral LD50 to rats is 50-60 mg/kg. It can be used as contact insecticide with some ovicidal properties. It is effective against aphids, capsids, leaf miners, thrips, sawflies. Also used for fumigation of glass houses. Marketed as 95% alkaloid or as nictotine sulphate (40% alkaloid) also as 3-5% dust.

(3) Ethylene dibromide

It is very heavy colourless and odourless liquid which is used as a soil fumigant to control wireworms. It is sold in solution with a petroleum oil and contains either 20 to 40% by weight. It is poisonous to man when inhaled as vapour or in contact with skin causing burning and irritation. It is inflammable so should be stored in light container in a cool place.

(4) Methyl bromide

It is a colourless gas which is heavier than air. Under pressure, it becomes a heavy colourless liquid. The gas is non- inflammable, almost odourless, very penetrating and poisonous to man. It can be used on a wide range of fruits and vegetables to destroy insects. It has been successfully used in fumigating cotton seeds to control the pink bollworms.

(5) Naphthalene

It has been used as an insecticide for controlling the cloth moths. It is also used as a repellant to combat the carrot rust fly and as a greenhouse fumigant. It is phytotoxic so it should be used only by experienced persons. An unrefined or crude naphthalene has been found effective as a soil fumigant for controlling wire worms. It is sprinkled along the sides of fresh furrows from top to bottom as the field is ploughed. It is white crystalline compound with a penetrating but not unpleasant odour. The fumes are very irritating to the eyes.

(6) Aluminium phosphide

It is a yellow crystalline solid, stable when dry but in the presence of moist air phosphine gas is librated which is inflammable in air. The phosphine gas is highly toxic and can be used for grain fumigation to control storage pests. Commercial product is a mixture of Aluminium phosphide and ammonium carbonate which is available as 3 gm tablet 20 mm diameter and 5 mm thick.

Note: Like chlorinated hydro carbon insecticides, a number of organophosphorus compounds, systemic insecticides and fumigants which were used in the past, are not in use now.

5. ANTIFEEDANTS

The chemicals which possess the property of inhibiting the feeding of insects are termed as antifeedants or antifeeding agents. These chemicals neither kill nor repel the insect but only inhibit feeding by making the food distasteful. A number of carbamates have antifeedant properties, in addition to the insecticidal action.

Many compounds e.g. nicotine will temporarily inhibit the feeding activities of insects, but in the case of antifeeding agents this inhibition is more or less permanent and is the main cause of death.

The fungicides : fentin acetate and fentin hydroxide have been shown to inhibit feeding in a wide range of plant feeders including *Prodenia litura* larvae and the adults and larvae of *Leptinotarsa decemlineata.* It is relevant to note that these compounds also exhibit chemosterilant activity and insecticidal toxicity against a number of pests. Some of the common antifeedants used for controlling a wide variety of insect pests like beetles, weevils, Lepidopterous larvae etc. are Triazene, Brestan, ZIP (a complex compound of zinc salt), Baygon, Pyrethrum.

6. ATTRACTANTS

The attractants are the materials used to lure insects into traps and poisoned baits. The use of geraniol in sugar syrup as ant baits and fermenting syrup for trapping various moths are good examples. Other substances like eugenol, sugar, molasses, yeast extracts, milk, sweet fruits, amyl acetate and ethyl alcohol are the common insect attractants. The natural or synthetic pheromones mixed with poison baits have also been used as attractants for the control of insects.

The insects recognize the food and oviposition sites by olfaction and to a certain extent communicate with each other by the use of specific odours, the pheromones produced by a variety of glands and comprise a range of chemical types. The only pheromones which have been exploited on a large scale for insect control are the sex attractants which bring about mature sexes together for mating purposes. The gypsy moth, *Porthetria dispar* is the only such species in which sex attractant was employed as a pest control method. The synthetic sex attractants i.e. pheromone analogue, named Gyplure was almost as active as Gyptol - the natural benzene extract of the abdomen of virgin females, but the Gyplure was much easier to synthesize. In field trials, quantities as small as 10^{-7} μg were able to attract males.

A great deal of success has been achieved, particularly with synthesized chemicals which are attractive to various species of fruitfly. One of the first compound to be found and used successfully to attract the Mediterranean fruitfly, *Ceratitis capitata* was Siglure. This fruitfly was also attracted and eliminated by another compound, Trimedlure mixed with insecticide. In a similar manner the Oriental fruitfly, *Dacus dorsalis* has been eliminated by methyl eugenol while a related compound eugenol (I part) mixed with geraniol (9 parts) has proved an effective lure for the Japanese beetle, *Popillia japonica.* Trimedlure and methyl eugenol have been combined with a third compound Cue-lure, which is an attractant for the melonfly, *Dacus cucurbitae.* In this way three of the most important fruit fly pests are being brought under control by the use of attractants. Likewise, anethol attracts the adults of the codling moth, *Cydia pomonella.* Butyl sorbate has been used to monitor the numbers of Chafer beetles.

7. REPELLENTS

These are the chemicals which repel or drive away the insects although they do not kill them, thus they have the effect of keeping insects away from the food sources. Insects exhibit a phenomenon of negative chemotropism hence they keep away from articles smeared with repelling agents. Different types of repellents are needed for different types of insects because all insects cannot be repelled by the same chemical.

Repellents must be safe, economic and not too unpleasant to use, their efficiency depending on their ability to deter the greatest number of species from feeding for the longest period. The duration of protection afforded by a chemical depends to a great extent on where it is applied e.g. if placed on the skin, it is liable to be diluted by perspiration, to be oxidized, hydrolyzed or volatilized or simply removed by abrasion. Thus, skin is usually protected for only a few hours but clothing may retain repellents for a few days or a week or more. As soon as repellents ceases to be potent enough to offset the attractive qualities of host, insect pests will be able to return to their normal activities with consequent lack of protection. Some of the important chemicals which drive away the insects are :

Bordeaux mixture : It is useful in driving away fleas, beetles, and leaf hoppers. It also destroys fungus and insect eggs. It is prepared with 2.5 kg of copper sulphate, 2.5 kg of hydrated lime and 250 litres of water. The copper sulphate may be dissolved in water in a glazed vessel and lime is diluted in another container. The inner water is then poured into copper sulphate and stirred vigorously. Finally the whole mixture may be diluted with the remaining quantity of water. The Bordeaux mixture renders sprayed plants distasteful and acts as a stomach poison for chewing insects, when mixed with 1/2 to I kg. of lead arsenate.

Creosote or coal tar : The creosote, coal tars and other derivatives of coal tar like naphthalene and carbolic acid etc. are successfully used as repellents to bugs, cloth moths, flies, beetles, and termites. These derivatives of coal tar can be mixed with soap and used as sprays.

A number of oils are used as repellents e.g. kerosene, cedarwood oil, fish-oil, castor oil, tar oil, pine oil, resin oil, citronella oil etc. are useful to protect the bodies of animals from insect attack. Similarly powdered alum, naphthalene, camphor, turpentine, calcium carbide, sulphur, tobacco dust, dry leaves of neem and powder of dry neem seeds can be used for dusting purposes. Benzyl and benzyl benzoate were found repelling to chigger and skin mites, dimethyl phthalate repells mosquitoes and mites and trichlorobenzene repells termites and ants.

The most widely used repellents these days are synergisted pyrethrins applied as sprays. In this case they act as repellents and not toxicant and their efficiency increased by the addition of compounds such as furfuryl alcohol (MGK 11). The compound furfuryl alcohol is also repellent to cockroaches and other household pests.

PRECAUTIONARY MEASURES NEEDED WHILE USING AND HANDLING THE INSECTICIDES

Before using any insecticide the operator must read carefully all the instructions given on the label of the packing which provides complete information about the nature of the poison, directions of their uses and warnings etc.

It is well known that most of the insecticides used now a days are poisonous to man and all other animal life including domestic animals. The insecticide enters the body by numerous ways and vitally influence the various systems of the man and other animals e.g. poisoning is caused by absorption through skin, inhaling fumes, or dusts, or by ingestion through several sources. (1) Hence care should be taken that no insecticide gets in the eyes, nose, mouth or skin. (2) They must be stored in safe places where no children can reach. (3) After the use of insecticides, the operator should wash hands etc. (4) Hands should not be dipped in any preparation of the insecticide and all the equipments used in operation should be washed. (5) Some of the insecticides may be absorbed through the skin. The preventive measures consists of using goggles, rubber gloves, washing of the contaminated cloths etc. (6) The hazards of inhaling dusts, mists and gases may be overcome by using respirators. (7) The cloths of the operator should be changed and the empty containers of the insecticides should be burried deep in the soil and away from the water sources. (8) When treating with fumigants, avoid exposure to heavy concentration or inhalation of gas for longer duration and use gas masks. (9) After the required exposure for fumigation, the room should be opened and entrance be allowed only after thorough airing.

10

INSECTICIDE APPLICATION EQUIPMENT

Broadly speaking the insecticide application equipments can be classified into (a) sprayers, (b) dusters and (c) granule applicators.

Fig. 31. Tractamount sprayer

(By Courtesy of American Spring and Pressing Works Ltd., Bombay)

(1) Sprayers : There are several types of sprayers (Figs. 31, 32A-D, 33 A,B,E) and they are mainly used for applying insecticides in liquid form. They may be either hand operated (hand sprayers) or electrically operated (electric sprayers) or operated by power (power sprayers). The most common hand sprayer is the hand atomizer, fine mist or spray can be obtained by this machine which carries a small tank at the base of a syringe and sends out a spray through a nozzle. Compression of air results in throwing out of the liquid from the end of suction tube extending into the tank with the blast of air breaking up the liquid into a mist form. This is the most widely used apparatus for domestic purposes, gardens and nurseries. Special nozzles are available for emitting a jet at right angles for spraying under leaf. Another common sprayer used for domestic purpose is the bucket pump sprayer. This equipment essentially consists of a long stout pump, a long delivery tube and an air chamber to keep reserve pressure during the upward stroke of the piston. This apparatus generally involves the presence of two operators one for handling the pump and the other for directing the spray.

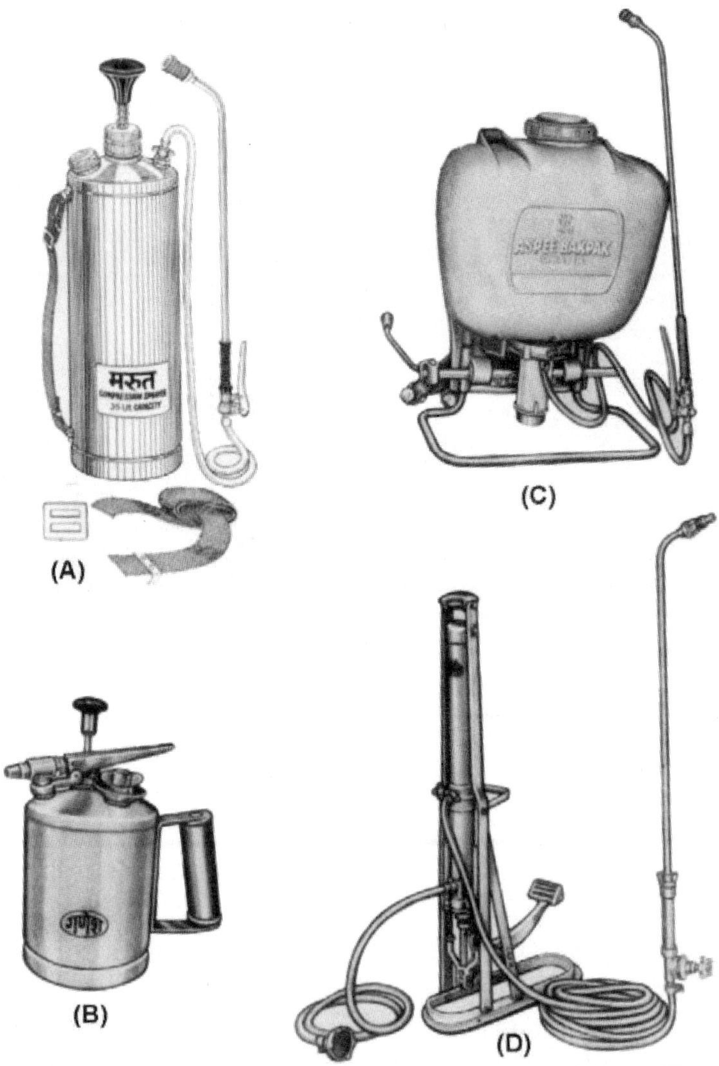

Fig. 32. A. Hand compression sprayer, **B.** Hand sprayer, **C.** Pesticide sprayer, **D.** Foot sprayer. (By courtesy of American Spring and Pressing Works Ltd., Bombay)

A more efficient and convenient hand sprayer is the Knapsack sprayer (Fig. 33A). This apparatus is a combination of a tank pump, delivery tube, nozzle and straps by which it can be hung over the shoulders. One of the hands of the operator works on the pump and the other directs the spray. This equipment serves very well for spraying insecticides in fruit gardens and small agricultural plots as well as hilly tracks for valuable crops such as tea and coffee.

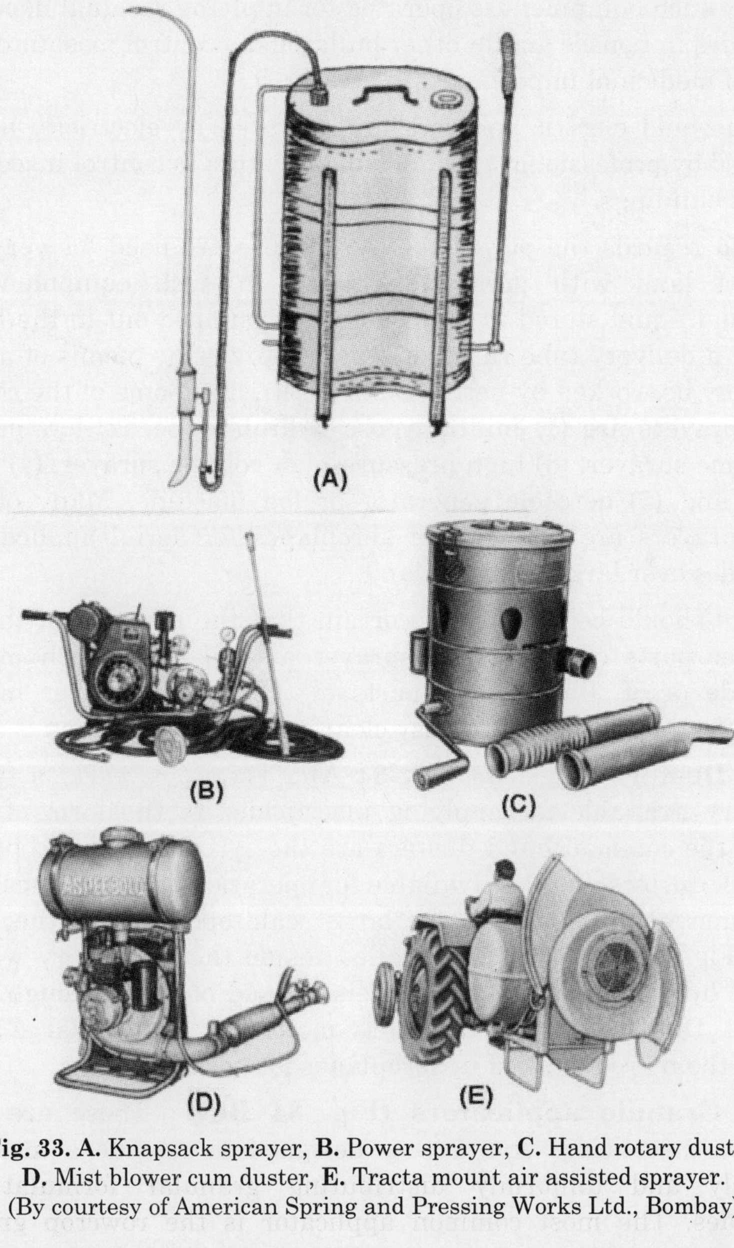

Fig. 33. A. Knapsack sprayer, **B.** Power sprayer, **C.** Hand rotary duster, **D.** Mist blower cum duster, **E.** Tracta mount air assisted sprayer. (By courtesy of American Spring and Pressing Works Ltd., Bombay)

For large scale operations the most convenient machinery is the compressed air sprayer. This equipment consists of a pump fitted in a tank which is operated by hand to generate considerable pressure inside the tank which in turn results in producing a continuous spray of the insecticidal liquid through the nozzle. The entire equipment can be handled by one operator. A pressure gauge is also fitted to this machinery to read the amount of pressure generated in the tank. Normally such equipment is operated for applying residual deposits of insecticides in houses and in other buildings to control mosquitoes and insects of medicinal importance.

Household electric sprayers are operated by electricity and are often used by professional pest control operators to control insect pests in public buildings.

With regards the power sprayers, these are used for very large tracks of land with great advantage. In such equipments the insecticidal liquid stored in a large tank is pumped out in the form of spray by a delivery tube through several nozzles by means of a motor which may be worked by petrol, oil or electricity. Some of the common power sprayers are (1) multipurpose hydrolic type, (2) low pressure low volume sprayer, (3) high pressure high volume sprayer, (4) blower sprayer and (5) aerosole generator or fog machine. Many of these power sprayers are installed in aeroplanes for aerial application of insecticides over large tracks of land.

Care should be taken to ascertain that the material of the tank and other parts of the power sprayer are all safe which must be insecticide proof. Besides the parts of the power sprayer must be detachable and should afford easy cleaning and replacement.

(2) Dusters (Figs. 33 CD, 34 A) : There are various types of machinery available for applying insecticides in the form of dusts. Some of the common hand dusters are the syringes or hand blowers. Knapsack dusters are also available for operation on a large scale. The most commonly used duster for large scale operation is the Rotary duster (Fig. 34 A). A fan is installed inside the machinery which is operated by one hand and this expels a cloud of dust through one or two exits. Dusters are also driven by motors for dusting large tracks. Many of them are installed in aeroplanes for aerial dusting.

(3) Granule applicators (Fig. 34 BC) : These are newly developed insecticidal equipments designed to provide a means for accurately and uniformly distributing granular formulation of insecticides. The most common applicator is the rowcrop granular applicator. This serves as a convenient tool for applying granular insecticides for control of insects on plants arranged in a row.

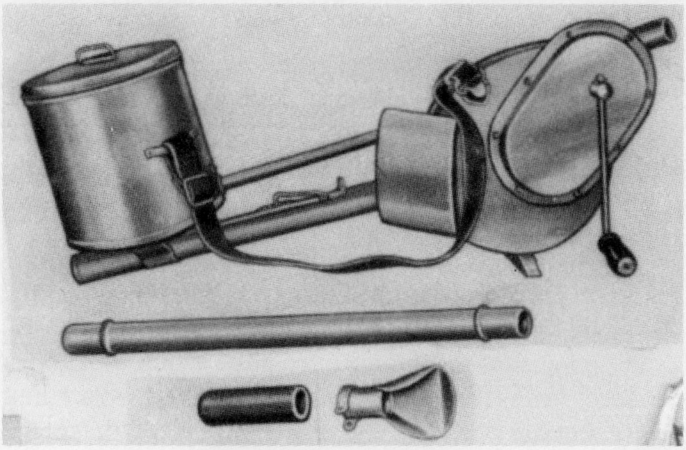

Fig. 34A. Rotary duster
(By courtesy of American Spring and Pressing Works Ltd., Bombay)

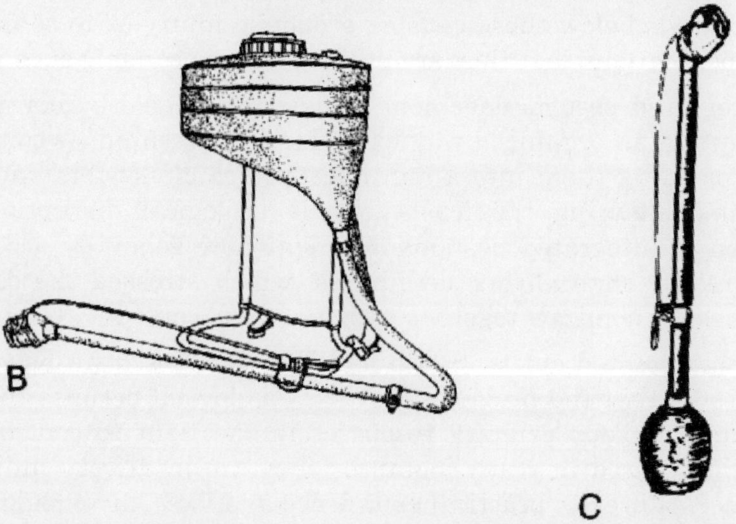

Fig. 34. B. Knapsack granule applicator, **C.** Hand granule applicator.
(By courtesy of American Spring and Pressing Works Ltd., Bombay)

Another type of granular applicator is the trailor type applicator widely used in large farms. They are equipped with metering devices for regulating the dosage.

Concurrent with the attempts to develop new insecticides the efforts and investigations are carried out by man to device novel and more modern type of insecticidal application equipments which may be in the form of sprayers, dusters and granular applicators.

11

INTEGRATED PEST MANAGEMENT

DEFINITION AND CONCEPT

A brief mention may be made here that when a single method of insect pest control fails, then several other methods are adopted at one time. According to Hill (1975) integrated control or the integrated pest management (IPM), is now defined as a pest population management-system that uses all suitable techniques (physical, cultural, chemical and biological) either to reduce the pest populations and maintain them at levels below those causing economic injury or to so manipulate the populations that they are prevented from causing such injury.

Integrated pest management achieves this ideal by hormonizing techniques in an organized way by making the techniques compatible and by blending them into a multifaceted flexible system and at the same time ensuring the least possible ecological disruption. This definition of integrated pest management now generally accepted is much broader than earlier definitions, which stressed the complimentary use of chemicals together with biological control.

It was pointed out by Smith and Van den Bosch (in Kilgore and Doutt, 1967) that the previous attitude was logical because these two approaches are our primary resources in the fight to control insect pests and as used in many instances they are in direct conflict with each other. However, practical experience and logic have made it clear that we must integrated not only just chemical and biological control but all available procedures and techniques into a single pattern aimed at profitable crop production and minimal environmental disturbance. Conway (1972) found that in Sabah, excessive use of pesticides was severaly aggravating the local pest situation and the most effective immediate measure was to curtail spray applications to allow survival of the predators and parasites which normally keep most of the pest populations in check. Rabb (1970) arrived at another difinition where he stated that originally integrated control tended to refer to the modification of insecticidal control in order to protect and enhance the activities of benificial insects (predators and parasites). Subsequently, integrated control interpretations have become more comprehensive until now some definitions of integrated control embody most of the essentials of pest management. Rabb preferred the term pest management because it connotes a broader ecological basis and a wider variety of opinions in devising the solutions to the pest problems.

According to Reay (1969) the need for an integrated system has arisen from the short comings of each method. Chemical insecticides are essentially drastic and short term in their effects and result in the deposition of toxic materials - on food stuffs and livestock. Moreover. they often have a broad spectrum of activity and can so alter the biotic components of the ecosystem as to cause the emergence of new problems. He further remarked that the biological agents on the other hand are safe to use but long term in action and may permit a pest to destroy a substantial part of a crop before they exert their effect. The correct use of integrated systems presupposes that a great deal is known about the behaviour of the organisms which are found in the larger area and involves the very careful use of insecticides. Only those compounds or formulations which will affect the pest more than the rest of the fauna can be used and their timing and placement must be related to the biology of any beneficial species present. Thus, Reay. (1969) adds that (e.g. the use of DDT) formulations in which the toxic particles are covered with a protective coating ensures that the compound will act only as a stomach poison to phytophogous insect pests and will not affect beneficial species as a contact toxicants. Similarly, the timing of insecticidal spray can ensure that parasites or predators will not be harmed because it is known that they are elsewhere or in a resistant state at that time. Even broad spectrum compounds may be used as long as they do not have prolonged residual activity and are employed as "spot" treatments or if sufficient areas are left untreated to maintain as reservoir of beneficial organisms. Reay (1969) has further mentioned that because of the lack of truly selective chemicals and the existence of what frequently amounts to scanty knowledge of the ecology of the pests and their enemies, integrated system have found only limited application. However, there are certain exceptions too.

Dent (1991) recognized three distinct phases in the development of an IPM programme (1) problem definition, (2) research, and (3) implementation (See chart) and suggested that the success of a

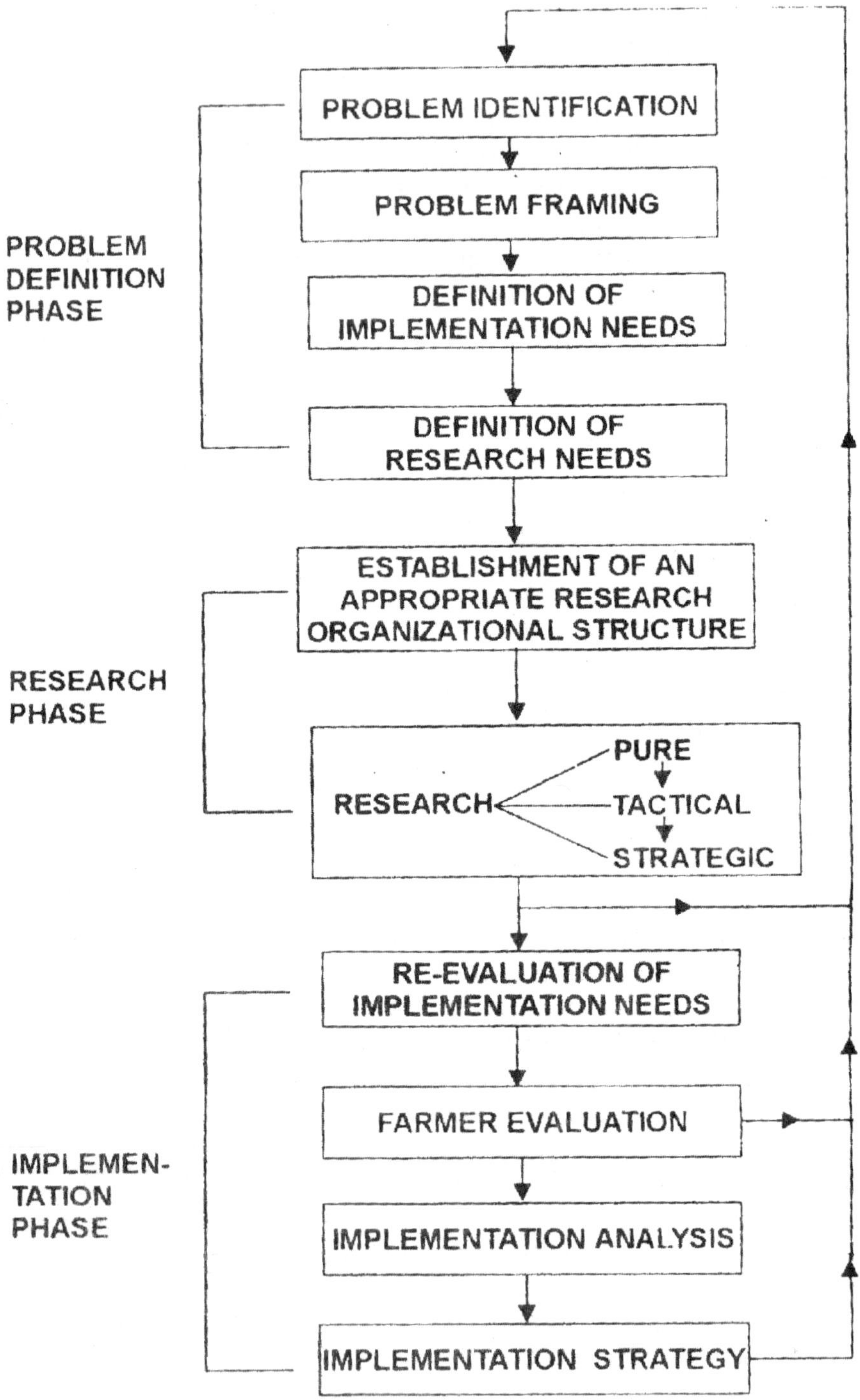

Chart showing for three phases in the development of an IPM programme (From Dent, 1991)

programme depends largely on the time and effort given to the first of these. Clearly, the more information that can be collated for the definition phase, the easier it will be to identify the research needs and implement the management strategy that will provide the maximum economic and socially acceptable returns. In addition to basic "scientific" information on a pest's ecology and current methods of control, there are important socioeconomic aspects to be considered; i.e., losses in yield must be placed in the context of the farming system in which they occur, and any changes in strategy must be formulated in consultation with, and meet with the approval of farmers. In many early projects, relatively little information was available on which to base IPM strategies, and these were developed largely by intuition, despite which they were remarkably successful. Now a days the considerably greater body of information generally available is stored in computers for use in constructing and testing mathematical models of the agroecosystem. Once developed, such models can be used to examine the effects of varying one or more of the factors that influence the population density of a pest and therefore, to determine the optimum method of reducing and maintaining this density below the economic injury threshold.

AGROECOSYSTEM

The following are some of the questions that might be asked in the development of an IPM strategy for the insect pests in a particular agroecosystem (remember that insects typically are only one component of the pest problem, and that any strategy for their management must be compatible with those developed for other pests and with overall socioeconomic goals). (1) What is the economic injury threshold? It has been observed frequently that insecticide users were maintaining pest population density at levels such below the economic injury threshold, increasing their operational costs in both the short and the long term. (2) What are the best parasitoids and predators to use and how might their abundance be increased? (3) Are some of the pests amenable to control by microbial insecticides? (4) What are the best insecticides to use, when should they be applied and what are suitable doses? These aspects should be considered not only for the pest complex but also with respect to the beneficial insects in the system and thus require knowledge of the specificity of action, stability, and possible side effects of the insecticides. (5) Are suitable cultural procedures available? (6) Are resistant strains of the crop available? (7) Do the farmers already have the resources, human and technological, to adopt the new strategy? (8) Will the new strategy be acceptable to farmers, both economically and socially? Culture, tradition, and religion are important factors in determining farming practice. Thus, changes in planting and harvesting dates may be resisted if they interfere with other customs. Likewise, a suggested change in the cultivars used may not be acceptable if the new cultivars' taste, texture, or yield is different. It will be apparent from this sample of questions that IPM is interdisciplinary in nature and may require the collaboration of experts in widely different areas, e.g., entomologists, plant breeders, chemists, toxicologists, meteorologists, computer modelers, economists, anthropologists, and sociologists. (Dent, 1991; Gillott, 1995).

Many examples of pest managment might be cited that differ in complexity, geographical location, and crop pest being controlled. In its simplest (and rarest) form, pest management involves only one method of control and, strictly speaking, cannot be described as IPM. Thus, biological control is entirely satisfactory for control of sugarcane pests in Hawaii and coconut pests in Fiji (DeBach and Rosen, 1991). More often, two or more control methods are employed. For example, at the height of the pesticide syndrome, apple growers in Nova Scotia used a battery of fungicides and insecticides (especially broad spectrum synthetics) against apple pests. Far from achieving the desired effects, this led to even greater problems. By about 1950 earlier pests such as codling moth (*Cydia pomonella*), eye spotted bud moth (*Spilonota ocellana*), Oystershell scale (*Lepidosaphes ulmi*) and fruit tree leafroller (*Archips argyrospilus*) had become pests because of destruction of their natural enemies. The need to reduce the escalating costs of this programme stimulated intensive study of apple orchard ecology, including the effects of pesticides on the natural enemy complex. The outcome of this work was the development of an integrated control programme that utilized fewer but more specific insecticides (as well as different fungicides, some of which had killed some natural enemies) and allowed for recovery of many of the natural enemies. By 1955 the scale insect had practically disappeared, and the density of the eye-spotted bud moth had fallen significantly.

As a final example, the integrated control programme for cotton pests in the San Joaquin Valley, California, will be outlined. This region produces about 10% of the cotton grown in the United States. As with many other crops, cotton growing was initially a small scale agricultural activity, not requiring an organized programme of crop protection. With the development of synthetic insecticides, a massive increase in acreage devoted to cotton growing occurred. Insect pest outbreaks were easily dealt with through the regular but infrequent application of an organochlorine insecticide. With time, however, it became necessary to increase the frequency and dose of insecticides, or to change over to using organophosphates or carbamates to combat population resurgences and increased resistance in the two major pests, the corn earworm (*Heliothis zea*) and a mirid bug (*Lygaeus hesperus*), as well as outbreaks of secondary pests such as cabbage looper (*Trichoplusia ni*), beet armyworm (*Spodoptera exigua*), salt-marsh caterpillar (*Estigmene acraea*), pink bollworm *(Pectinophora gossypiella*), tobacco budworm (*Heliothis virescens*), and spider mites (Tetranychidae) following destruction of their natural enemies. At the peak of this ``crisis phase", almost 50% of all insecticides applied to field crops across the United States were being used on cotton, and control often required 10 to 20 applications of insecticide per growing season (Gillott, 1995).

Early in the development of the integrated programme, it was realized that no reliably established economic injury thresholds existed for either of the major pests and that, for *Lygaeus,* time of insecticide application was critical. For *Lygaeus* it was determined that the generally accepted injury threshold of 10 bugs/50 net sweeps was invalid; rather, the major period of importance was June 1 to July 20 and only if the density exceeded 10 bugs/50 net sweeps in this period should insecticides be applied. After July 20, *Lygaeus* densitites of 20 or more bugs/ 50 net sweeps would not affect the quantity or quality of cotton produced, provided that the plants had flowered normally in June and early July.

ECONOMIC THRESHOLD

It was noted that corn earworms seldom became pests except in fields treated for *Lygaeus* outbreaks (using the old value for economic injury threshold) where the corn earworms natural enemies (the bugs *Geocoris pallens* and *Nabis americoferus*. and the lacewing, *Chrysopa carnea* were destroyed by insecticide. With the new threshold for economic injury due to *Lygaeus*, tied in with the plant's seasonal development, the use of insecticides was reduced, especially after late July, and the earworms natural enemies survived. Further reduction in insecticide usage came about with the raising of the economic injury threshold for corn earworms from 4 earworms/plant to 15 treatable earworms (first-and second instar larvae) per plant. This arose, in part, from realization that larvae often feed on surplus buds, flowers, and small bolls that are not picked.

Cultural control is also used to reduce the effects of *Lygaeus* on the cotton crop. *Lygaeus* prefers alfalfa (*Medicago sativa*) to cotton and will remain on the former plant when given a choice. Traditionally, however, alfalfa fields were solid cut (all of the field is cut at once) provoking mass migration of *Lygaeus* into adjacent cotton fields. Alfalfa growers were persuaded to practice strip cutting to prevent such a migration, particularly following the discovery that strip cutting also reduces pest problems in the alfalfa itself by favoring *Aphidius smithi*, a parasitoid of the pea aphid (*Acyrthosiphion pisum*). An earlier proposal that cotton growers plant strips of alfalfa through their fields has not become practical because of difficulty in managing both crops. It was also noted that safflower could be a major source of *Lygaeus*, the bugs moving into cotton as the former plant began to flower. Careful monitoring of the bug populations in the safflower, and a single insecticide treatment when 70% of the insects are third - to fifth instar nymphs, largely eliminated the threat to adjacent cotton fields. Monitoring by means of light traps and pheromone traps (using "gossyplure", the latter have been particularly valuable for assaying pink bollworm movements into the San Joaquin Valley) is now also routinely employed to determine the relative abundance of adult pest Lepidoptera (Dent, 1991, Gillott, 1995).

Another cultural method that may yet become commercially feasible is the use of food sprays in attempts to prolong the seasonal effectiveness of natural enemies whose populations frequently decline after mid season. For example, in experimental plots adult *Chrysopa* were attracted to the food sprays and produced more eggs, resulting in significantly less injury to plants by corn earworms.

Microbial control has become an important method for use against some lepidopterous cotton pests. For example, *Heliothis* nuclear polyhedrosis virus against bollworm and tobacco budworm in California and other states in lieu of broad spectrum insecticides.

Integrated control results in substantial savings, both financial and in terms of pollution, through the greatly decreased use of insecticides. For example, in the San Joaquin Valley, the traditional insecticide - based control programme for cotton pests cost about twice as much per acre as the integrated programme, and even greater savings may be incurred as the programme is further improved. From 1976 to 1982, the average quantity of insecticide applied per acre to cotton decreased by 72% for the United States as a whole. This, in combination with a decrease from 60 to 36% in the proportion of acres treated, resulted in a drop in cotton's share of the field crop insecticide market from 49 to 24%.

Despite these remarkable statistics, many crop growers and government officials remain convinced that the unilateral use of insecticides is best. Persuading these individuals that to continue with this approach is not only economically unsound but will lead to long-term, perhaps irreversible deterioration of environmental quality is perhaps the greatest challenge that scientists have ever faced (Gillott, 1995).

12

ECOLOGICAL TACTICS FOR MANAGEMENT OF INSECT PESTS

In practice, the integrated pest management (see Chap. 11) requires a stepwise decision-making process, beginning with definition of management goals or desired future conditions (1) Defined goals determine which available management options are most appropriate in a given situation. (2) Adequate estimates of population size and the predicted time to reach the action threshold establish the timeframe for initiating necessary action. (3) Evaluation and comparison of available management options (including no action) on the basis of expected accomplishment of management goals, effectiveness within the established timeframe, cost, envrionment and social consequences, etc. permits ranking of options in terms of their net benefits relative to costs (Gatto *et al.,* 2009). In many cases, insect control may not be the best option, e.g., if compensatory growth prior to resource harvest would replace lost production. Computerized decision-support models have been developed to aid the evaluation of management options. Such models integrate a user interface with submodels that predict population growth, effects of management actions and calculated benefit/cost to provide benefit/cost analysis under various user defined scenarios. Finally, monitoring of management activity and evaluation of the benefits and costs of the completed project are essential to improved future (adaptive) management. Ecological tactics available for managing crop and forest pests, are described below, which are based on a comprehensive work of Schowalter (2011).

ECOLOGICAL TACTICS FOR MANAGING CROP AND FOREST PESTS

Management of crop and forest pests has been a major application of insect ecology for the past two centuries (see: Sschowalter, 2011). Forests present special challenges for pest management. Access to tall canopies for assessment of injury levels is limited. Aerial surveys can provide data on severity and extent of injury, but ground checking is necessary to confirm the identity of the pest. Further, forests are subject to the cumulative effects of insects and pathogens, as well as other factors, that predispose forests to other factors for long periods before injury becomes apparent. This requires a degree of attention to changes in forest conditions that often is not supported within operating budgets. Forests in many regions are managed for multiple uses that may be affected by insects in complementary ways. Finally, controlling insects in forests often is not cost-effective, given marginal profits and losses due to pest control or unscheduled salvage harvests.

A number of remedial options are available when the suppression of pest populations becomes necessary. Modern insecticides, including a variety of insect growth regulators (IGRs, such as methoprene) and chitin synthesis inhibitors (CSIs, such as noviflumuron), remain an option when rapid reduction in pest populations becomes necessary. Their use can be minimized by restricting application to trap crop borders around crop fields (Butler *et al.,* 2009) and by employing a variety of ecologically based alternatives.

Many viral (e.g., nuclear polyhedrosis virus, NPV), bacterial (e.g., *Bacillus thuringiensis,* Bt), and fungal (e.g., *Beauvaria*) pathogens can be applied in the same manner as chemical controls and often are confused with chemical insecticides. The advantage of microbial control agents is their ability to replicate within their hosts and spread epizootically through the target population.

The effectiveness of many microbial pathogens has led to genetic modification of some crops to express toxic bacterial proteins, such as Bt endotoxins Cry1A, Cry1Ac and Cry-2Ab. Although controversial in many countries, transgenic corn, wheat, cotton and other crops have reduced the use of more toxic insecticides substantially and show fewer effects on non-target species than do insecticides. Nevertheless, pollen and plant detritus containing Bt toxins may affect soil organisms and decomposition, and they may be exported to other ecosystems, such as streams, where they could affect non-target organisms.

Widespread planting of transgenic crops in some regions threatens to undermine their long-term effectiveness, given the ability of insects to adapt quickly to any strong selection factor. Therefore, a high dose with refuge strategy is recommended to prevent survival of pests on the Bt crop and to maintain a large, non-adapted population in non-Bt refuges (Schowalter, 2011).

Incipient resistance in some target species has led to insertion of multiple Bt toxin genes into some crop plants to delay insect adaptation, to the extent that cross-resistance to the different genes is minimized. Nevertheless, long term effectivenss could be undermined if either or both Bt toxins are not produced at concentrations sufficient to kill all herbivores, or if pollen contamination of non-transgenic refuges or native vegetation leads to variable Bt concentrations across the landscape. This requires attention to soil moisture, nutrient availability, or other conditions that stress crop plants and influence production of Bt toxins, and to the landscape structure of Bt and non-Bt crops (especially for insects with broad host ranges that might include multiple transgenic crops), as well as cooperation among scientists, growers and government agencies.

Other genetic tools to control pest populations are being explored, but these also may have unintended consequences for communities and ecosystems. RNA interference (RNAi) is a transgenic method that can inhibit target insect expression of P-450 genes for detoxification of plant defenses. RNAi is accomplished by incorporating double-stranded RNA (dsRNA) specific to the detoxification gene. It was demonstrated that adding dsRNA to the diet of cotton bollworm, *Helicoverpa armigera,* decreased the expression of CYP6AE14, the bollworm's primary detoxification gene, in the midgut and reduced larval growth. This technology could permit focused control of a specific target species in specific situations with minimal non-target effects (Burt, 2003, See: Schowalter, 2011).

Homing endonuclease technology (Burt, 2003) involves engineering a homing endonuclease gene (HEG) that encodes for an enzyme that recognizes and cleaves a specific 20-30 bp sequence on chromosomes that do not contain the HEG. The intact HEG+ homologue is used by the cell's repair system as the template, so after repair both chromosomes contain the HEG. The gene would spread quickly through the population from a relatively small number of introduced individuals, providing for rapid manipulation of population characteristics. For example, mosquitoes could be engineered to make them ineffective as disease vectors. Although such options seem attractive because of their apparent focus on the target species, they fail to address potential effects of the engineered population on the network of direct and indirect interactions in ecosystems, e.g., the variety of vertebrates that interact with mosquitoes.

A promising new tool is application of chemical elicitors, such as jasmonic acid, to induce production of natural defenses by crop plants in advance of herbivory. This approach would be most effective when infestations can be anticipated reliably, and action thresholds are high, because of delay in expression of induced defenses. However, expression of natural or transgenic defenses by plants depends on adequate resources for growth and defense.

Pheromones have become highly valued tools for pest management, because of their specificity in attracting target species. Identification of the particular blend of volatile compounds that are used by a species is necessary for best effect. Pheromones can be used in a number of ways for management purposes. Pheromone-baited traps can be placed in strategic locations to detect and monitor the distribution and abundance of a target species, such as a potential invasive species. Pheromones in controlled release containers placed in the field or infused in light, plastic filaments applied aerially can be used to disrupt the normal attraction of mates or to repel insects from the treated site. Baited traps also can be used to reduce target populations locally. These applications will be most effective in the control of small or relatively isolated populations, for which saturation of the populated area with pheromone is feasible, and attraction of additional target insects from outside the patch is limited, e.g., orchards and islands (Yamanaka and Liebhold, 2009).

Sterile insect release (SIR) is an option in some circumstances (Yamanaka and Liebhold, 2009). This technique requires mass rearing, sterilization and inundative release of sterile insects to saturate the target population with non-reproductive individuals. A critical requirement for success is a single-mating population. If females can mate more than once, unsuccessful mating with a sterile mate does not necessarily prevent reproduction. Eradication of the screw worm fly, *Cochliomyia hominivorax,* from the U.S. during the 1960s, and subsequently from Mexico by 1990, remains a classic example of the successful use of this technique, but required the release of millions of sterile males at a total cost of US$750 million. Other sterile insect release programms have been most successful on islands (Yamanaka and Liebhold, 2009).

Advances in remote sensing and geographic information systems (GIS) have improved the precision of pest management activities. Within-field patterns of pest density can be superimposed on crop production gradients to identify areas of greatest potential increase in yield under different pest management scenarios. In place of traditional broadcast methods, pest management can be focused at locations where it will be most economically or ecologically beneficial. Furthermore, fertilizer and pesticide application rates can be tailored to habitat conditions to avoid over-application and minimize the contamination of groundwater or streams (see Schowalter, 2011).

13

MOLECULAR TECHNIQUES AND HABITAT MANIPULATION FOR PARASITOID CONSERVATION IN CROPPING SYSTEMS

The need to manage insect pests has always been a key component in agricultural production systems. At the beginning of the twenty-first century, different strategies for pest management coexist and are sometime perceived as contradictory by the industry, farmers and consumers. On one hand, low-input and organic farmers base their pest management programms on the integration of a series of cultural, biological and ecological tactics aimed at reducing pest survivorship and/or impact. On the other hand, high-input conventional farmers rely on pesticides, mechanical control practices and transgenic crops to reduce pest impact. A transgenic plant is defined as one that has had foreign genetic material purposefully introduced and stably incorporated into the plant genome through means other than those that naturally occur in the environment (Shatters, 2000), i.e. transgenic plants are those which carry additional stably integrated and expressed foreign gene(s), usually transform from unrelated organisms. The whole process of introduction, integration, and expression of the foreign gene(s) in the host is called genetic transformation or transgenesis (Chahal and Gosal, 2002). Both approaches to pest management are reflected in the current scientific literature. In the following pages a brief account of molecular techniques for conservation in cropping systems has been given which is based on a comprehensive work of Fabian *et al.* (2004) with minor changes in the text material so as to suit the graduate student of entomology.

In recent years, several researchers have evaluated the potential of habitat manipulation and conservation of biological control to promote natural enemies of agricultural pests (See: Barbosa, 1998). The goal of these practices is to manipulate the environment to enhance the survivorship and/or performance of natural enemies.

Alongwith the growing interest in ecological approaches, there has been an expansion of research into use of transgenic organisms as a component of high input conventional farming systems. The debate about ecological risks and benefits of using transgenic organisms to combat arthropod pests is very vast. Further, the potential negative outcomes include evolution of resistant pests, impact on non-target insects, escape of transposable genes into wild relatives, cascading multitrophic level impacts, and changes in farm management practices that could reduce the survivorship of natural enemies. But the potential benefits include high efficiency in pest population regulation, reduced environmental impact of pesticide applications, and increased crop yield.

The contributions of molecular technologies to pest management go beyond the development and release of transgenic organisms. Among the areas that could benefit are an increased ability to identify pests and natural enemies, evaluation of failure or success in the establishment of natural enemies, study of dispersal patterns, and assessment of genetic divergence among and within populations. Despite these advantages, few field ecologists have fully incorporated molecular tools into their research.

MOLECULAR TECHNIQUES AND PARASITOIDS

Although parasitoids represent a valuable and diverse group of natural enemies of agricultural pests, their success in annual cropping systems has been limited. Among the factors constraining the effectiveness of parasitoids as biological control agents are their small size and their high species diversity that could hinder correct identification. Further, the harsh environment created by agricultural practices reduces parasitoid survivorship.

Parasitoids constitute an ubiquitous and diverse group of insects with enormous ecological and economic importance, particularly as biological control agents of agricultural pests. However, the small size of many parasitoids and their high species diversity have hampered their correct identification and selection for pest-management programms. The molecular methods can increase our ability to characterize parasitoid biology, genetic diversity and ecological requirements. This in turn allows an unambiguous identification of parasitoid species and an improved selection of the best genotype to be released as a biological control agent. Molecular methods can also help field entomologists better characterize the parasitoid community already present in agricultural systems, understand host associations and assess the evolution of parasitoid behavioural traits that affect host selection. These factors could play important roles in the success of parasitoids as biological control agents. The following is a brief introduction to some of the molecular tools used to solve questions related to parasitoid ecology, biology and population genetics. For a detailed description of molecular methods see Fabian *et al.* (2004).

(1) Enzyme electrophoresis

The allozymes are variant proteins produced by allelic forms of the same locus that can be separated by electrophoresis. The analysis of variation in allozymes (a reflection of changes in their gene codes) was the earliest molecular method successfully used to determine genetic variation among insect species and for taxa identification. Loxdale and Hollander (1989) and Menken and Ulenberg (1987) provide a detailed summary of cases in which allozyme electrophoresis has been used in agricultural entomology. In the specific case of parasitoids, this technique allowed successful discrimination between several species of *Trichogramma* (Hymenoptera : Trichogrammatidae), a genus of minute parasitoids of great importance in biological control. More recently, the use of allozymes has been supplemented with techniques generating direct or, indirect estimates of nucleic acid variation.

Although enzyme electrophoresis is inexpensive and relatively easy to conduct, it investigates only some of the variation in the most conserved class of DNA (the slowly evolving coding DNA), underestimating the amount of genetic variation in the non-coding DNA. This non-coding DNA, also called `Junk', parasitic or `selfish' DNA, may constitute 30-90% of the insect genome (Hoy 1994). In addition, insects in the order Hymenoptera (most of the effective parasitoids used in agriculture) have exceptionally low allozyme variability.

DNA molecular markers and nucleotide polymorphisms

The recent advent of molecular methods to directly investigate the DNA molecule has increased accuracy and resolution in genome analysis. DNA molecules are constructed of monomeric units called nucleotides consisting of a purine or pyrimidine base, a pentose and a phosphoric acid group. Molecular methods allow direct evaluation of genomes, determining the proportion of nucleotide sites differing between two or more DNA sequences (nucleotide polymorphisms). Genomes can also be compared indirectly by assessing the bands produced in an electrophoresis gel by DNA fragments obtained from known individuals (DNA molecular markers).

The number of molecular techniques available for entomological studies has greatly expanded since the advent of the polymerase chain reaction (PCR). Kary Mullins conceived the PCR method in 1983, receiving the Nobel Prize in chemistry for the invention. PCR is an *in vitro* method for amplifying a target region of DNA by means of enzymes that catalyze the formation of DNA from deoxyribonucleoside triphosphates, using single-stranded DNA as a template. The presence of conserved regions in DNA sequences such as the mitochondrial DNA and nuclear ribosomal DNA makes it possible to amplify fragments from organisms for which there is no specific sequence information available. Although the part of the genome to be used will depend on the level of resolution needed by the researcher, several mitochondrial (COI, COII, 16S) and nuclear (ITS1, ITS2, 28S, D2, D3 and EF-1) regions have proved useful for discriminating parasitoid species (see: Fabian *et al.,* 2004).

The number of PCR-based techniques is expanding every year and they have been successfully used in ecological and entomological studies. Among the different areas that have benefited from PCR-based techniques are systematics, population genetics and insecticide resistance assessment. PCR-based techniques offer several advantages, including the possibility of working with extremely small insects, such as many of the effective parasitoids used in biological control. They are also unaffected by the life stage of insects and can potentially be used with stored, dry or old material. Currently, the main disadvantage of using molecular methods is the cost involved.

Some of the PCR-based techniques commonly used are (1) restriction fragment length polymorphism (RFLP) analysis (2) microsatellite analysis (3) single-strand conformation polymorphisms (SSCPs), (4) random amplified polymorphic DNA (RAPD) and (5) amplified fragment length polymorphism (AFLP) fingerprinting. Table 1 presents a comparison of different molecular methods in terms of their sensitivity, cost, efficiency, level of discrimination and application.

The selection of any given molecular technique depends on the type of problem to be solved the cost of the technique, its ease of use and the sample size to be analyzed. The section of the genome to be investigated depends on the level of variation that the researcher needs to resolve. Among the different PCR-based techniques, RAPD-PCR has been very popular in entomological studies perhaps because the technique does not require prior knowledge of DNA sequence, provides a rapid way of identifying genetic markers, is inexpensive and is easy to develop. However, RAPD markers have been criticized as a tool for molecular identification of species because their results can have poor reproducibility. Reineke *et al.* (1999) recommended amplifying the DNA in a second reaction in order to assess the reproducibility of the banding patterns produced by the RAPD primers used in a test.

Table 1 : Molecular tools for assessing variation in DNA fragments of PCR products and qualitative assessment of the cost, effectiveness, levels of discrimination, type of data obtained and applications of the different tools.

Attribute	PCR-RFLPs	Microsatellites	SSCPs	RAPD-PCR	AFLP
Sensitivity	Moderate	High	Moderate-high	Moderate	High
Cost	Low	Moderate	Low	Low	Moderate
Efficiency	High	High	High	High	High
Level of discrimination type of data (gene frequencies or base pair changes)	Single nucleotides in nuclear and mt-DNA/gene frequency data	Differences in individuals and populations in nuclear DNA/ neither gene frequency nor base pair changes	Single nucleotide in nuclear DNA/gene frequency data	Differences in single nucleotides in nuclear DNA/gene frequency data	Differences in single nucleotides in nuclear DNA/gene frequency
Application	Diversity, geographic variation, hybrid zones and species boundaries, phylogeny	Mating systems, diversity, parentage, relatedness	Parentage, relatedness	Hybrid zones and species boundaries, biotype and subspecies identification	Subspecies identification

MOLECULAR TECHNIQUES USED IN BIOLOGICAL CONTROL

(1) Cryptc species identification

The unequivocal taxonomical identification of pests and natural enemies is important in biological control. Although evaluation of morphological features is the predominant tool to distinguish insect species, many parasitic groups are visually undistinguishable (cryptic species) despite their high genetic diversity. This genetic diversity, in turn, could be important to the success of parasitoids as biological control agents.

Beginning in 1994, two *Ageniaspis* (Hymenoptera : Encyrtidae) populations, one from Australia and one from Taiwan, were introduced into Florida as part of a classical biological control project against the Asian citrus leafminer *Phyllocnistis citrella* (Lepidoptera : Gracillariidae). Although taxonomic specialists identified the two populations as belonging to the same species, *Ageniaspis citricola,* there wre clear biological and physiological differences between them. By 1996, the Australian population of *A. citricola* had colonized most of Florida's 344 250 ha of citrus groves and parasitism of citrus leafminer pupae was found to be as high as 99% in some sites. Alvarez and Hoy (2002) collected *Ageniaspis* individuals in 10 field sites at six Florida countries and took advantage of RAPDs to show no evidence that the Taiwan population had established in Florida by the end of 1999. After these results were obtained, only Australian *A. citricola* was sent to Caribbean and Latin American countries for biological control of the citrus leafminer, to ensure establishment.

(2) Differentiation between exotic and indigenous parasitoids

In biological control programms where exotic strains of parasitoids are introduced into a new area, it is not always possible to distinguish between imported and native strains using morphological features. Molecular methods have been used to differentiate strains of indigenous and exotic biological control agents prior to their release. For example, many strains of exotic parasitic Hymenoptera that are investigated in the US to control the Russian wheat aphid, *Diuraphis noxia* (Homoptera : Aphididae), are indistinguishable from the indigenous ones. Narang *et al.* (1994) took advantage of RAPD-PCR to identify and characterize four species of hymenopteran parasitoid introduced for the control of the Russian wheat aphid and develop an easy to use dichotomous key for parasitoid identification. Zhu and Greenstone (1999), using the resulting PCR product in a gel of the amplified ribosomal ITS2 region, were able to distinguish three species and two strains of *Aphelinus* (Hymenoptera : Aphelinidae) endoparasitoids that had been released against the Russian wheat aphid. In South Africa, Prinsloo *et al.,* (2002) used specific primers for both the ribosomal ITS2 and mitochondrial 16s DNA sequences of several morphologically similar parasitoid species and strains of *Aphelinus*, digested them with a restriction enzyme and showed in an agarose gel electrophoresis the differences between strains. These procedures allowed a reliable separation of *Aphelinus* species and confirmation that the exotic aphid parasitoid *A. hordei* had successfully spread in wheat fields, suggesting it could be effective at controlling the Russian wheat aphid.

(3) Parasitoid rearing

The contamination problems are recurrent in laboratory colonies of parasitoids, especially when rearing small, morphologically similar species. In many cases, the biological control worker is not aware of the contamination but molecular methods can be used to monitor the taxonomic quality of parasitoid colonies or to assist in selecting desirable traits prior to release. For example, Rosen and DeBach (1973) cited that cultures of *Encarsia perniciosi* (Hymenoptera : Aphelinidae) imported from Germany to control the San Jose scale, *Quadraspidiotus perniciosus* (Homoptera : Diaspididae) were contaminated with a related but ineffective species, *E. fasciata.* It is estimated that about 4 million *E. fasciata* were released, but never became established.

There is fairly extensive prior work on the use of molecular markers for identification and characterization of *Trichogramma* species and strains. For example, a number of workers (see Fabian *et al.,* 2004) have found specific markers for several morphologically similar *Trichogramma* strains and species with the use of RAPD, ITS2 sequences and RFLP analyses. These DNA markers could potentially be used in rearing laboratories for rapid species identification.

AGRICULTURAL PRACTICES AND PARASITOIDS

The modern agricultural practices create a harsh environment for parasitoids. High-input conventional annual cropping systems usually depend on large amounts of off-farm chemical and mechanical inputs such as insecticide and herbicide applications, primary and secondary tillage, fertilization, cultivation and harvest. From an ecological point of view, these management practices represent disturbances that disrupt the characteristics of the ecosystem, the community or the population through a rapid change in resources, substrate availability or physical environment.

On a temporal dimension, annual crop fields can be characterized as ephemeral and predictable habitats where early successional plants (the crops) are established as short-lived monocultures. These plants have been selected to have low chemical and mechanical defences and are readily attacked by herbivores. Spatially, the current tendency of removing non-crop habitats has severely reduced and fragmented native habitats that could provide resouces for parasitoids including over wintering sites, pollen and nectar, suitable environmental microsites and alternative hosts. In this way, industrialized agriculture generates a mosaic of small and isolated late-successional patches with high species diversity embedded in a matrix of extensive early successional agricultural fields with low species diversity.

The joint impact of distrubances, temporal predictability, resource allocation and habitat fragmentation has been identified as a limiting factor conditioning the abundance and effectiveness of parasitoids in annual agroecosystems. These unique biotic and abiotic characteristics, coupled with the relatively low cost and high effectiveness of pesticide applications, has discouraged the use of natural enemies, including parasitoids, as a tool to manage arthropod pests in annual cropping systems.

IMPROVING PARASITOID SURVIVORSHIP THROUGH HABITAT MANIPULATION AND MOLECULAR BIOLOGY

The habitat manipulation, defined as a series of environment manipulations to provide natural enemies with the necessary resources to improve their effectiveness at combating agricultural pests, emerges as a unifying theme to mitigate the negative impact of agricultural practices. The underlying concept of habitat manipulation is that provision of supplementary and complementary food, microclimate modification and existence of refuge habitats in close association with crop fields might help natural enemies to cope with the detrimental impact of agricultural practices. For example, many species of adult parasitoids used wildflowers and aphid honeydew as food resources that are not provided in agricultural fields in which weeds are controlled. Practices such as no tillage and conservation tillage, cover cropping, crop residue conservation, intercropping and establishment of herbaceous strips in close spatial association with crop fields represent viable within-crop field approaches to provide the necessary resources to enhance parasitoid survival.

Further, as explained in the following pages the molecular tools can be integrated with an in-depth knowledge of insect biological and ecological requirements to help entomologists improve habitat manipulation practices for parasitoid conservation. The integration of these tools and knoweledge could be done at different spatial scales.

Within - crop field activities and molecular based techniques

Pesticide applications can be the most important factors reducing parasitoid survivorship within crop fields. In the short term, pesticides can kill large numbers of parasitoids, which may lead to an outbreak of secondary pests. In the long term, repeated pesticide applications may select resistant biotypes and influence parasitoid population dynamics. Ruberson *et al.* (1998) described different approaches to integrate pesticide applications with natural enemies. Among the recommended tactics are periodic scouting of crop fields, use of selective pesticides and sublethal doses, and spatial and temporal separation of pesticides and natural enemies.

The molecular technologies can help biological control practitioners to integrate pesticide use with natural enemies through the selection of pesticide-resistant strains of natural enemies. Although this approach appears to be more successful with predators than with parasitoids, recent progress in parasitoid selection programms could reverse this situation. For example, RAPD-PCR has proved to be a useful technique to distinguish between two populations of the walnut aphid parasitoid, *Trioxys pallidus* (Hymenoptera : Aphididae), differing in resistance to pesticides (Edwards and Hoy, 1995). Although the experiments were performed in the laboratory, the authors suggest that RAPD-PCR could be used to determine the fate of pesticide-resistant parasitoids released in the field.

Conserving parasitoids at the farm and landscape level

While practices at the within - field level might enhance parasitoid survivorship, agricultural practices occurring at larger scales might negate such conservation efforts. Several studies have shown that field size, crop rotation and presence of non-crop habitats play a critical role in determining within-field parasitoid abundance. For example, Landis and Haas (1992) determined that parasitism of the European cornborer, *Ostrinia nubilalis* (Lepidoptera : Pyralididae) by its larval parasitoid, *Eriborus terebrans* (Hymenoptera: Ichneumonidae) was significantly higher at the borders of maize (also known as corn) fields than in field interiors. They further determined that the greatest levels of parasitism were observed at wooded field edges. In accordance, it has been observed that access to plant nectar, aphid honeydew or sugar improved *E. terebrans* survival in crop and non crop habitats. These studies suggested that the absence of food resources in large maize fields, combined with high temperatures before canopy closure, was responsible for the low abundance of parasitoids observed in field interiors.

It has been determined that short rotation of periodical cutting of hedges and clover/grass grazing areas spread among crop fields forming a series of annual, biannual and semi-perennial habitats correlated with an increase in parasitism of the pest aphid *Sitobion avenae* (Homoptera : Aphididae). These and other studies highlight the importance of semi-natural habitats established in close association with crop fields for natural enemy conservation. Ideally, agricultural landscapes should contain a series of non-crop habitats interspersed with crop fields to provide shelter and food for parasitoids. Mid- and late-successional habitats provide parasitoids with adult food resources, alternative hosts, overwintering sites and shelter for adverse conditions. It is interesting to note that the parasitoids use semiochemicals in host location, some of which emanate from plants non crop vegetation interspersed with crop fields can also affect parasitoid behaviour. Understanding changes in parasitoid behaviour to plant assemblage composition is a key element in the development of habitat practices aimed at increasing vegetational diversity of agricultural systems. Yet due to their small size, vagility and the inherent difficulty of mark and recapture experiments, very little is known about parasitoid dispersal behaviour.

The molecular-based techniques may aid entomologists to assess the impact of agricultural landscape structure on parasitoid population and community structure. Vaughn and Antolin (1998) used RAPD-PCR markers visualized by SSCP analysis to study the population structure of *Diaretiella rapae* (Hymenoptera : Braconidae), a parasitoid of several aphid species. The authors found that larger genetic variation occurred between fields separated by short distances than between areas separated by longer distances. They also argued that the reduced genetic exchange among subpopulations indicates that released *D. rapae* would not disperse between fields (see Gurr *et al.,* 2004).

Althoff and Thompson (2001) used the mtDNA cytochrome oxidase 1 (CO1) gene sequence and nuclear rDNA RFLPs to compare patterns of host search behaviour among six populations of *Agathis* n. sp. (Braconidae : Agathidinae) located in a relatively large geographic area in south eastern Washington, USA. The results showed no isolation by distance, suggesting long distance dispersal among populations. The authors argued that phenotypic differences in host parasitoid interactions such as time allocated to searching, ovipositor length and place of searching appear to be driven by local plant characteristics. Although this study was not developed within the context of annual cropping systems, it provides a framework that could be used to foster the understanding of how local, mid- and large-scale agricultural landscape affect parasitoid abundance, searching behaviour and distribution (For review See : Fabian *et al.,* 2004).

CONCLUDING REMARKS

Despite the large potential of biological control in sustainable agriculture, the commercial development and success of this approach has been extremely limited. This is particularly true in annual cropping systems where disturbances, resource paucity and habitat simplification reduce natural enemy survivorship. More successful biological control may be posible if agricultural entomologists take advantage of newly developed concepts and techniques deriving from a wide variety of areas ranging from molecular based technologies to landscape ecology.

The molecular technologies are among the fastest growing sectors in modern agriculture with the potential of transforming farming systems. To date, much of the debate on this issue has been on the benefits and risks of releasing transgenic organisms. As discussed, these newly developed technologies can also increase the adoption and success of biological control agents through greater taxonomic accuracy, higher precision in insect identification and a better understanding of the genetic and population structure of natural enemy populations.

There is a rich collection of evidence that ecological diversity plays a key role in the resilience of farm systems. More specifically, increasing diversity appears to be a cornerstone of habitat manipulation to enhance parasitoid survivorship.

Further, encouraging diversity however, might not be enough and sometimes it may even have negative consequences. Instead, it is necessary to identify the key elements of diversity needed for the particular set of parasitoid species under consideration. Undoubtedly, correct insect identification is crucial. Parasitoids and alternative hosts must be accurately identified before any existing information on their environmental needs could be used in the selection of within field, farm and landscape level practices that will enhance their survivorship and effectiveness. Molecular techniques provide a useful tool to advance the accuracy and ease of parasitoid and host identification (See: Fabian *et al.,* 2004).

14

POLYPHAGOUS PESTS

1. WHITE ANTS

TERMITES (Isoptera: Termitidae) (Figs. 35, 36)

These are commonly called white-ants, the name being used because though not related to ants, they live in colonies and in many of their ways resembles the ants. Their colour varies from whitish pale to brown. Termites form the most important insect pest as they cause very great damage. The termites are polyphagous. Their food is principally composed of cellulose which they obtain from living or dead vegetation, books and other articles of wood, fabric etc.

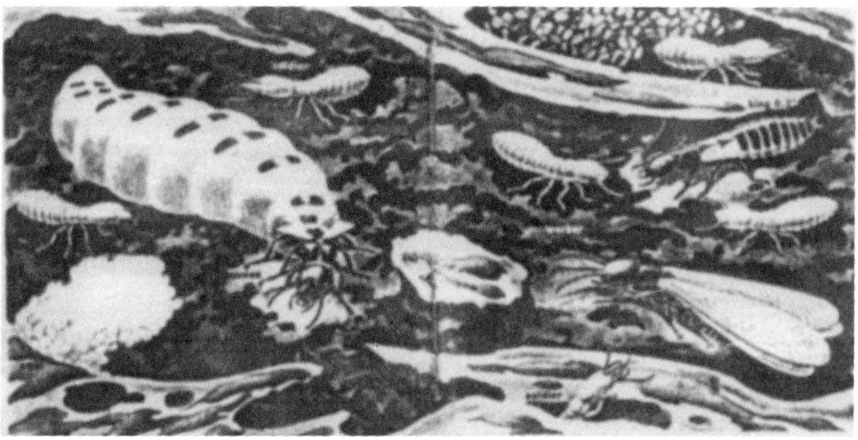

Fig. 35. Colony of termite.

Maize, jowar, bajra, pulses, paddy, wheat, sugarcane, potato, tea, coffee, coconut, jute, sunnhemp, fruit trees, forest trees are badly damaged by termites. They cause considerable damage to the dry and dead wood, furniture made out of the dry wood, books and almost all articles containing cellulose.

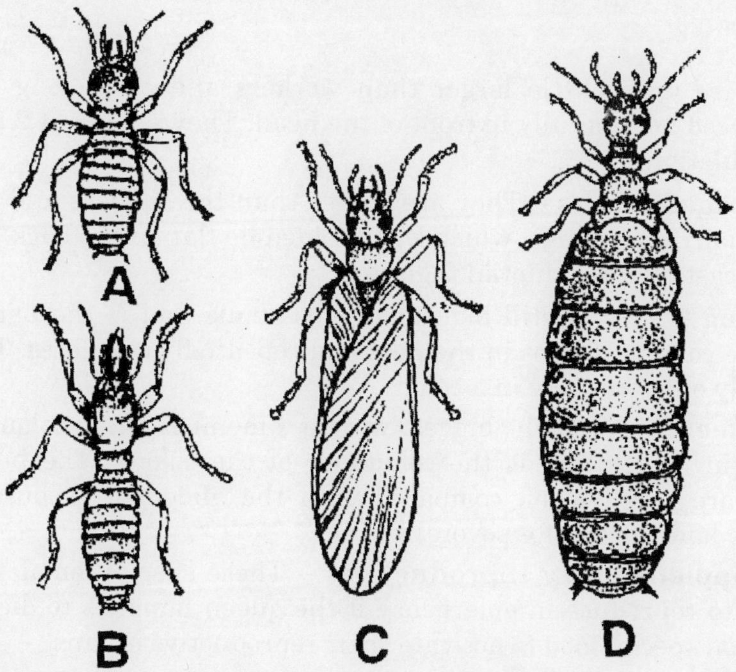

Fig. 36. Castes of termite. **A.** Worker, **B.** Soldier, **C.** King, **D.** Queen

Life history

The termites have two main forms viz. reproductive and sterile. The colony consists of one male or king, one female or queen and a variable but generally large number of workers and soldiers. The queen is a giant and has no other work than of continuously laying millions of yellowish or cream coloured kidney shaped small eggs. They are deposited in particular cells for hatching. The nymphs resemble miniature adults. The nymphs undergo several moults producing workers and soldiers. Periodically they produce winged forms or primary reproductives, which swarm out and repeat the cycle. With the first few showers of rain there particularly appear a few winged male and female forms. They leave the termitarium when mature, fly about, copulate, shed their wings become kings and queens, burrow in soil and establish new colonies.

Workers

These are small white insects most commonly seen doing damage. They can be easily recognized by their vertically carried head and small broad jaws. These are injurious form and constitute 80-90% of the whole population.

Soldiers

These are a little larger than workers and carry long pointed jaws placed horizontally in front of the head. They are about 2 to 3% of the population.

Winged adults – They are bigger than the soldiers and possess four equally long wings, which when folded lie flat on the back and are readily cast off after nuptial flights.

King – This is still bigger wingless male and is the partner to form the colony. It lives in the nest and repeatedly copulates. There is generally only one king in a colony.

Queen – This is the biggest wingless member of the colony about 10 cm. in length and is the cofounder of the colony. The head and thorax are small when compared with the abdomen. Thousands of eggs are laid per day for several years.

Supplementary reproductives – These special females whose duty is to reproduce in emergency if the queen happens to die. These are fed on special food to activate their reproductive organs.

Control – The affected parts in the trees should be sprayed with paris green or sanitary fluid or osfoetoid solution banding the trees by coal tar, renewed half yearly.

Protection of wooden structures can best be obtained by taking care that no wood comes in contact with the ground and that light and dryness are ensured as much as possible. Timber thoroughly treated with coal tar, creosote are repellant to the insects.

When termites are already in the wood, all infested parts will generally need to be removed and the nest should be located if possible and destroyed by using kerosene or by fumigation. Treatment of the soil near a building with a mixture of trichlorobenzene one part, fuel oil or creosote two parts aids for a few years in keeping termites away.

Termites penetrate through lime mortar but not cement provided there are no cracks. A plinth 8-15 cm. thick of cement before erecting walls will bring about desired safety. The floor of the house may be treated with arsenic solution.

Soil treatment for eradication or as a protective barrier must be thoroughly applied. Around the foundation, the normal procedure is to trench the soil and flood or rod-inject the formulation into the soil. The rate for shallow foundation is 9 litres per lineal 1.5 m. or deep foundation 10 litres per lineal 1.5 m.

All wood that have contact with the soil should be termite proof by pressure impregnation with coal tar, creosote etc. When building, no bits of waste wood such as grade stakes and concrete forms should be left burried in the soil near the structure. Stumps and waste wood should be removed from the premises.

Soil treatment with the dust of endosulfan 4% @ 25-30% kg/ha be done uniformly after the last ploughing and before sowing the seeds.

2. DESERT LOCUST

Schistocerca gregaria **Forskall (Orthoptera: Acrididae) (Fig. 37)**

It is a polyphagous pest and virtually all crops are at risk so far as locusts are concerned but there is some preference of cereals. The leaves and soft shoots are eaten from margin inwards leaving irregularly shaped feeding marks. Swarm damage usually results in complete defoliation of crops.

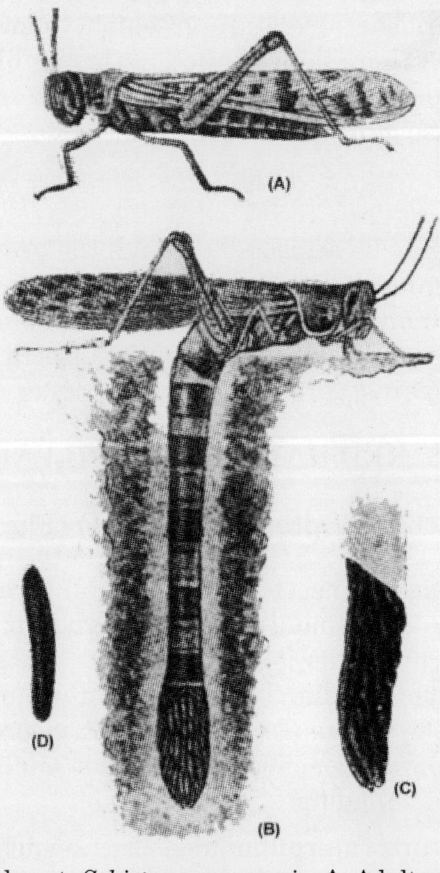

Fig. 37. Desert locust, *Schistocerca gregaria.* **A.** Adult gregarious phase, **B.** Egg-laying, **C.** Egg cluster, **D.** Egg (magnified) (from Pruthi, 1969)

Life history

The life history of a locust consist of egg stage, hopper stage and the adult stage. The eggs are laid in a hole made by thrusting the ovipositor 10-15 cm. deep in sand embedded in a frothy mass which hardens to form a tubular egg pod. Each egg looks like a grain of rice which is yellowish in colour and 7 to 8 mm. in length. Each egg pod contains 70-150 eggs and each female can lay 4-5 pods of eggs. Egg development takes 2-4 weeks according to temperature.

The first larva is rather vermiform and it has to wriggle up through the egg pod to the sand surface. On reaching the surface it moults and becomes a hopper over a period of 3-10 weeks depending on the temperature. The hoppers develop through 5 instars before they become adult. The hopper exist in two distinct phases, solitary and gregarious with difference in colour and behaviour.

The adults are large; male 40-50 mm. and female 50-60 mm. pale yellow or brownish. The elytra are greenish yellow, translucent with many brown spots. There is a characteristic peg like process between the bases of the forelegs and the under surface of the thorax is not smooth. There are several generations per year.

Control

The hoppers may be attracted through the use of suitable baits as well as dusting, ground spraying and aerial spraying with carbaryl and methyl parathion. Against aerial swarms, malathion is widely used. Barrier spraying with persistent insecticides can be very successfully used against large swarms.

3. RED HAIRY CATERPILLAR

Amsacta species **(Lepidoptera: Arctiidae) (Fig. 38)**

It is a polyphagous insect which feeds on a variety of plants like jowar, bajra, maize etc. Among the whole group of hairy caterpillars, the "red hairy caterpillar" (Kutra/Katra) is most injurious to agriculture throughout India. The habits and nature of damage etc. of all the species of the genus *Amsacta* (e.g. *A. moorei* Butl., *A. collaris* Hamp. and *A. albistriga* Walk.) are so similar that separate descriptions are not called for.

The fully mature caterpillar measures about 4-6 cm. in length. The body of the caterpillar is covered with reddish brown or brownish orange hairs. The moths are robust having white wings with black spots. The outer margins of the fore wings, the anterior margin of thorax and the abdomen is scarlet red. There are black bands and dots on the abdomen.

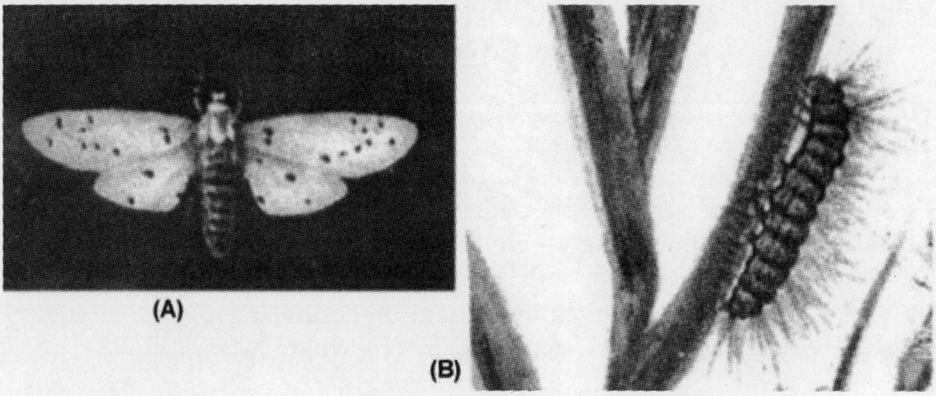

Fig. 38. Red hairy caterpillar *Amsacta* species, **A.** Adult, **B.** Caterpillar (From Fletcher, 1914).

Life History

The caterpillars of *Amsacta* are active from late June to September. The insect is normally univoltine and occasionally bivoltine; when there are two generations a year, the first generation is non-diapausing and the pupae of the second generation undergoes diapause.

The moths of the red-hairy caterpillars appear in large number after the first shower of monsoon. They lay light yellow, spherical eggs in large prominent clusters on the cultivated plants as well as on weeds. A single female may lay upto 1500 or more eggs. The larvae hatch out in about 3-5 days and begin to feed. They feed continuously, grow and migrate from one field to another field after almost completely browsing the area they visit. The caterpillars undergo 6-7 moultings and complete their development in 4-6 weeks depending on the climatic conditions. The full grown caterpillars go down into the soil for pupation. In the soil they remain for about 8-10 months i.e. whole of the winter and summer and emerge as moths after the first few good showers of rain. The duration of pupal period largely depends on the arrival of monsoon. There is only one generation in a year.

Control

The moths can be attracted by light traps. The clusters of eggs are quite large in size and that can be easily destroyed by hand picking. The young caterpillars of the *Amsacta* can be killed by dusting the cultivated crops with malathion 5% @ 30 kg/ha. For full grown caterpillars spraying endosulfan 35EC @1.5 litres in 1000 litres of water per hectare controls the pest.

4. SURFACE GRASSHOPPER

Chrotogonus trachypterus Blanch. (Orthoptera : Acrididae) (Fig. 39)

The surface grasshopper is world wide in distribution. The grasshoppers are polyphagous and besides cotton and wheat, they also feed on a number of other cultivated crops. Both the nymphs as well as the adults feed on leaves by cutting germinating plants of cotton, wheat etc. particularly in areas near by the waste lands.

Fig. 39. Surface grasshopper, *Chrotogonus trachypterus*

Life history

Chrotogonus remains active throughout the year but its activity is reduced during winter when only eggs and adults are found. A female may lay 36-434 eggs in 2-15 egg pods at a depth of 3-6 cm in the soil. Each egg pod consists of 16-46 eggs. In the summer, eggs hatch in 19-33 days where as in winter they hatch in about 5 months. The nymphs feed on grasses or other soft plants, moult 5-7 times and are full fed in 40-70 days. The adults are stoutly built, about 20 mm. long and 8 mm. broad with a rough greyish appearance. There are two longtudinal rows of black dots on the under side of the body which is otherwise white. The body of the female is larger and flat than the males; the females live for about 3 months and the male almost half as long. There are two generations in a year.

Control

The pest can be controlled by dusting the crops with malathion 5% or 12.5 Kg. 5% carbaryl (sevin) dust per hectare on first appearance of the grasshopper(s).

5. BIHAR HAIRY CATERPILLAR

Diacrisia obliqua **Walk. (Lepidoptera : Arctiidae) (Fig. 40)**

The Bihar hairy caterpillar is a sporadic pest and is widely distributed throughout the country particularly in Bihar, M.P., U.P. and some other states. It is a polyphagous pest, specially of sesumum, moong, linseed, mustard some vegetables and a number of other vegetable crops. The damage is caused by the caterpillars.

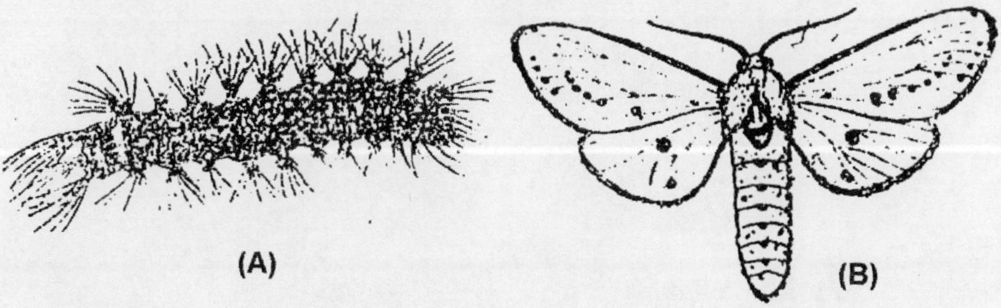

Fig. 40. Bihar hairy caterpillar, *Diacrisia obliqua:* **A.** Caterpillar, **B.** Adult

Life history

The pest breeds from March-April, and July-November. It passes the hottest part of summer and the coldest winter in the pupal stage amidst plant debris. The female lays about 400-1150 light green spherical eggs in groups of 50-100 on the under side of the leaves. The eggs hatch in about 4-10 days and the larvae feed gregariously. They grow to maturity through 7 stages within 3-6 weeks. The full grown caterpillars measure about 40-45 mm. in length. They are covered with long greyish hairs. The pupation take place in the plant debris or soil in a loose silken cocoon. The pupal stage lasts 1-2 weeks and the moth lives for about one week. The moth measures about 50 mm across the wings when spread. The head, thorax and under side of the abdomen is light yellow in colour. The antennae and eyes are black. The whole life cycle is completed in about 6-12 weeks and the pest passes through 3-4 generations in a year depending on the climatic conditions.

Control

The pest can be controlled by spraying the endosulfan 35EC or sevin W.P. 50% @ 1 kg/ha in 1000 litres of water or quinalphos 25EC @ 750 ml/ha mixed in 1000 litres of water.

15

PESTS OF COTTON AND FIBRE CROPS

1. RED COTTON BUG

****Dysdercus koenigii* Fabr. (Heteroptera: Pyrrhocoridae) (Fig. 41)**

It is a serious pest of cotton and sucks juice with its long beak from the seed of either green or ripe cotton bolls. Both the nymphs and adults are voracious feeders and cause extensive damage to the plant. The main symptoms of attack by this pest is the shrivelling of bolls in the plant. A severe infestation by this pest causes serious impairment to the vitality of the plant and destroys the germinating power of the seed by the loss of its oil contents. Besides these features, the formation of lint is badly affected.

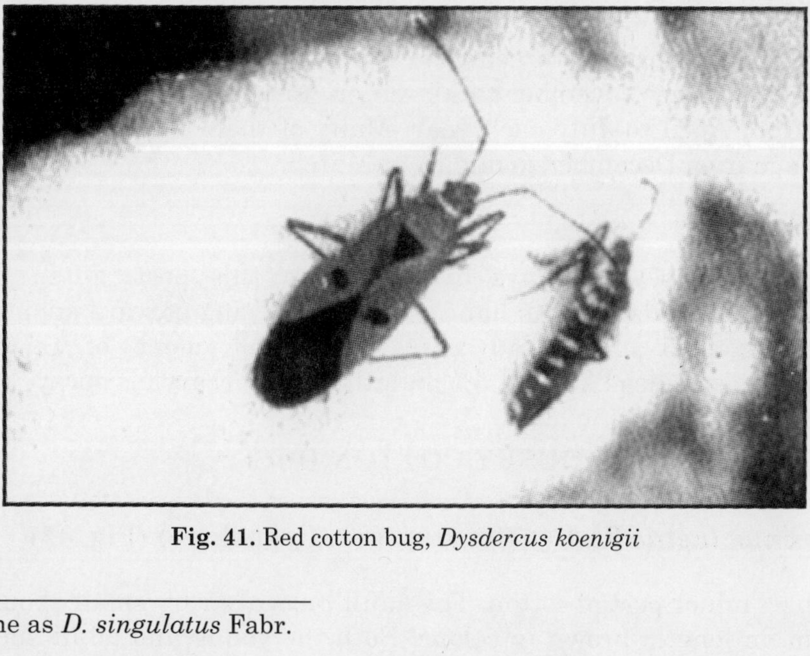

Fig. 41. Red cotton bug, *Dysdercus koenigii*

* Same as *D. singulatus* Fabr.

Although D. *koenigii* is a major pest of cotton, this insect has been reported to thrive well on other malvaceous plants like Hollyhock, lady's finger (Okra) or in other nonmalvaceous plants like sweet potato, tobacco etc.

Life history

The eggs of *D. koenigii* when freshly laid are soft to touch, they are almost round yellowish-white in colour and a little over 1 mm. long. They are laid in loose clusters, about 80-100 per cluster normally in cracks on the surface of the moist soil or under fallen leaves round about the plants.

The eggs hatch out under favourable conditions in about 7 days. Small, active yellowish-red first instar nymphs hatch out which resemble the adults except in size and absence of wings. They are about 1.5 mm. long dorsally flattened and convex possessing 4 – jointed antennae and two jointed tarsi. The nymph generally cluster round the cotton bolls. In course of time moulting occurs and there are normally 4 moults or 5 instars before a fully developed adult individual is formed. The average time taken for all the 4 moults is 30-35 days under favourable conditions. The duration between any two successive moults is dependent on many factors one of them being food. Under unfavourable conditions, there may be 5 moults and 6 instars and the moulting period may extend to 82 days. Generally the entire life cycle of this pest is completed in about 42-45 days.

This insect normally breeds on cotton from August to November, on Hollyhock from November to March or April, and on lady's finger (Okra) from April to July each year. Many of them hibernate in the adult stage from December to middle of March.

Control

There are several ways of controlling this pest. Often the mechanical methods such as hand picking, trapping etc. are adopted. Chemical control chiefly advocates the employment of certain synthetic insecticides like dust 5% malathion or carbaryl as spray.

2. DUSKY COTTON BUG

Oxycarenus laetus Kirby (Heteroptera: Lygaeidae) (Fig. 42)

It is a minor pest of cotton. The adult bugs are very small about 4 to 5 mm. in length, brown in colour. Both the young and adult sucks the sap of immature seeds of the cotton. Lint is stained when they are crushed in factories. There are several other host plants belonging to Family Malvaceae.

Fig. 42. Dusky cotton bug, *Oxycarenus laetus*

Life history

Each female lays white 15 to 40 eggs singly or in groups of 4 to 18 on the half open bolls which become pink later on. After about one week or more nymph emerges. There are seven instars. The total time taken for development ranges from 30 to 40 days. There are several generation in a year.

Control

The dusky cotton bugs can be controlled by applying methyl demeton 25EC @ 1 litre/ha in 1000 litres of water in the form of spray on the crop.

3. COTTON LEAF ROLLER

***Sylepta derogata* Fabr. (Lepidoptera: Pyraustidae) (Fig. 43)**

The cotton leaf roller occurs in all the cotton- growing tracks and is an important sporadic pest. It is distributed throughout the India. Damage is done by the caterpillars. The larvae feed on cotton leaves and in the years of serious outbreaks, cotton plants may be completely defoliated.

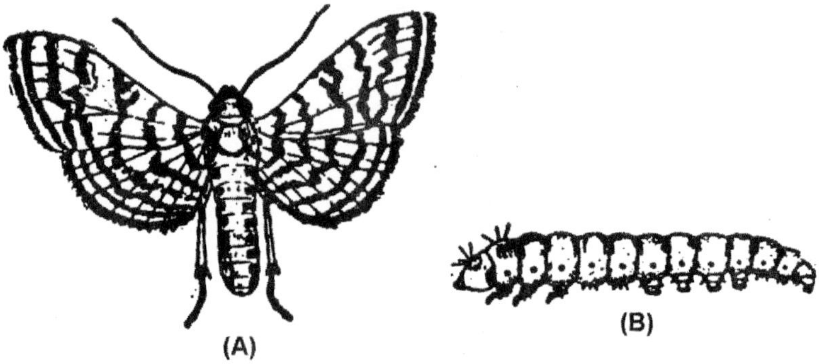

Fig. 43. The cotton leaf-rollar, *Sylepta derogata* **A.** Moth, **B.** Caterpillar.

Life history

The moths are active at night when they mate and lay 200-300 eggs singly on the under side of the leaves. The eggs hatch in 2-6 days and the young caterpillars at first feed on the lower surface of the leaves. The damage is caused by caterpillars. The larvae grow through 7 stages and are full fed in 15-35 days. The full grown larvae measure about 25-30 mm. in length. They are greenish-grey or pink and are found feeding inside the rolled leaves. Pupation takes place either on the plant or inside the rolled leaves or in the soil. They emerge as moths in 6-12 days and live for about a week. Moths are yellowish-white in colour with black or brown spots on the head and thorax. They measure about 28-40 mm. across the spread wings. They remain active from March to October. The life cycle is completed in 23-53 days and the pest passes through 5 or 6 generations in a year.

Control

The pest, as in other cases, can be controlled by spraying the infested crop with carbaryl 0.1% or endosulfan 0.05% or fenitrothion 0.075%.

4. COTTON JASSID

**Empoasca devastans* Dist. (Homoptera: Cicadellidae) (Fig. 44)

The jassids are popularly known as 'leaf hoppers', because their legs are specially developed and modified for hopping like those of locusts and grasshoppers. All the nymphal stages as well as adults inflict the same type of damage. Like other bugs, they suck out the cell

* Same as *Amrasca devastans* (Dist.)

sap from the plant tissue. Adults are about 3 mm. long and greenish yellow in the summer and have a reddish tinge in the winter.

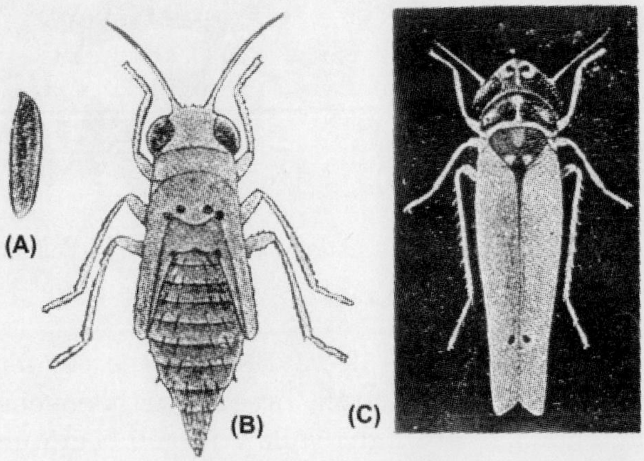

Fig. 44. Cotton jassid, *Empoasca devastans* **A.** Egg, **B.** Fifth instar nymph, **C.** Adult (From Pruthi, 1969).

Life history

The eggs are usually inserted into the parenchymatous tissue between the vascular bundles and the epidermis. Each female lays about 15 yellowish eggs, which hatch in 4-11 days. The nymphs moult 5 times and the whole life cycle is completed in about 2 weeks to more than a month and half depending on the climatic conditions. On transformation into winged adults, they live for 5-7 weeks feeding constantly on the plant juice. The pest completes 7 generations in a year but as many as 11 generations have been recorded in some parts of the country.

Control

The jassids can be controlled by spraying the infested crop at 15 days intervals with a number of insecticides like phosphamidon 0.02%, malathion 0.05%, endosulfan 0.05% and monocrotophos 0.03%.

5. COTTON WHITEFLY

Bemisia tabaci **Genn. (Homoptera: Aleurodidae*) (Fig. 45)**

The common name whitefly is applied to a group of small (about 1 mm. or so) sucking insects. The damage is caused by sucking the cell

* Also written as Aleyrodidae.

sap. The growth of the plants is adversely affected and when the attack appears late in the season, the yield is lowered considerably.

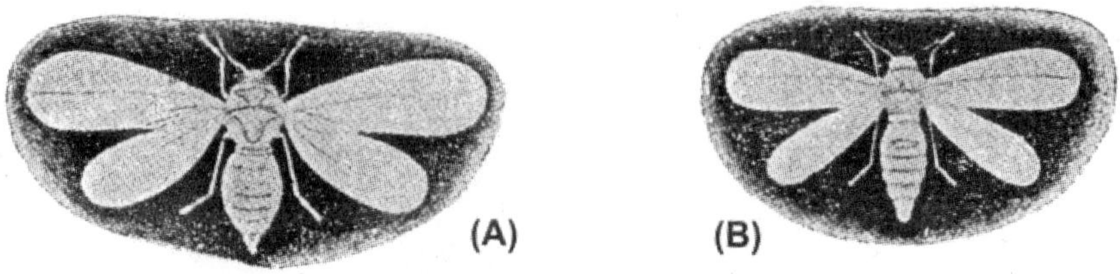

Fig. 45. Cotton white-fly *Bemisia tabaci* **A.** Male, **B.** Female

Life history

The females lay eggs singly from 70-119 eggs per female in 1-4 weeks time. The insect can often breed parthenogenetically i.e. without mating. The eggs are light yellow in the beginning but turn brown later on. They hatch in 3-5 days in April to September, 5-17 days in October – November and 33 days in December – January. The elliptical nymphs feed on the cell sap and grow into 3 stages to form the pupae within 9-14 days in April – September, and 17- 81 days in October- March. In 2-8 days, the pupae change into white flies. The whole life cycle is completed in 14-122 days and 11 generations are completed in one year.

Control

The pest can be controlled by spraying the infested crop at 15 days intervals with any one of these insecticides: phosphamidon 0.02%, malathion 0.05%, endosulfan 0.05%, methyl demeton 0.025%.

6. SPOTTED BOLL – WORM

**Earias fabia* Stoll (Lepidoptera: Noctuidae) (Fig. 46)

It is equally a serious pest of cotton and the young larvae devour the tender shoots, leaves, flower buds and ripening bolls of the plant. The lint and seeds are completely destroyed and the yield is reduced by 10 – 50%. Like *D. koenigii* this pest also infests other food plants like lady's finger (Okra), Hollyhock etc.

Life history

The female lays round eggs about 2 – 4 hundred in number singly on bolls, terminal leaves or buds or flowers. These eggs hatch in 4 – 7

* Same as *Earias insulana* (Boisd.)

days into dark pale greenish larvae with black dots and short spine like hairs on either side of the back. As soon as they hatch, they bore into the tender shoots and later on reach the flowers and bolls. The full grown larvae having a brownish white body, spotted with different colours like brown, green, black and orange. Hence it is called "spotted boll worm". The larval period lasts for 9–16 days depending upon atmospheric conditions. At the end of the larval period they move out of the boll and pupate in a tough grey silken cocoon on plant or soil. The pupal period ranges from 10-12 days after which fresh moth emerges. The entire life cycle spreads over a period of 26-35 days.

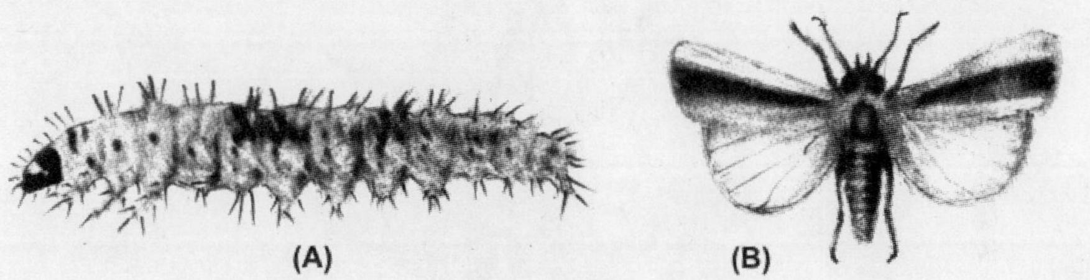

Fig. 46. The spotted boll worm *Earias fabia*. **A.** Larva, **B.** Moth (From Lefroy, 1909)

Control

While the mechanical control is occasionally employed by hand picking the caterpillars and destroying them, the most common method of control is spraying of the insecticides like endosulfan 35EC @ 1.25-1.50 litres/ha in 1000 litres of water, 0.1% carbaryl or 0.03% monocrotophos at 15 days intervals 2-3 time controls the pests. Removal of damaged parts of the plant is also useful.

7. PINK BOLL – WORM

***Pectinophora gossypiella* Saund. (Lepidoptera: Gelechiidae) (Figs. 47, 48)**

This is another serious pest of cotton, and the caterpillar 8-10 mm in length, damages the bolls vehemently, consequently the seeds inside the bolls are completely devoured and the formation of lint is affected. In fact the infestation of a boll commences with the tiny caterpillar entering the boll at its base through a minute hole. Inside the boll the caterpillar excavate a long sinuous tunnel eating away tissues on its ways and damaging the lint badly. It finally reaches to seed which it tunnels and feeds on the nutrient available in the seeds.

* Same as *Platyedra gossypiella* Saund.

It should be pointed out here that the caterpillars of this pest unlike those of *Earias fabia* never attack the shoots. In addition to cotton, lady's finger (Okra), Hollyhock etc. also serve as suitable hosts for this pest.

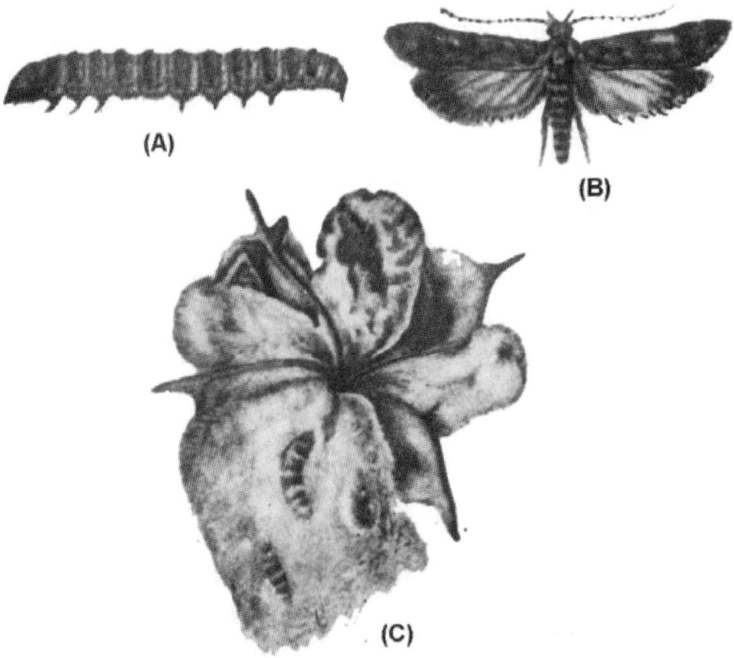

Fig. 47. Pink boll worm *Pectinophora gossypiella*, **A.** Larva, **B.** Adult, **C.** Infested cotton boll (From Fletcher, 1914)

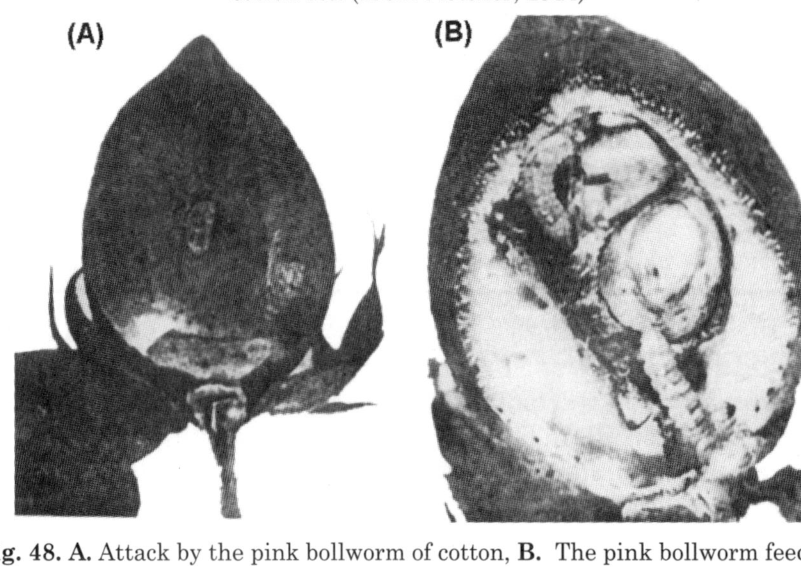

Fig. 48. A. Attack by the pink bollworm of cotton, **B.** The pink bollworm feeding inside a boll (From Atwal, 1976).

The adult moth is dark brown in colour 8-9 mm spread across the wings, with irregular black margins on the forewings. The hind wings are silvery grey and both wings are fringed with long hairs posteriorly. The tips of hind wings are sharply pointed.

Life history

The female lays elongate, oval eggs about 0.5 mm long and 0.25 mm. broad. They have minute longitudinal zigzag ridges and they are laid either singly or in groups on or near a bud or boll. The eggs which are green at first turn yellowish brown gradually and hatch in about 4-10 days into small active yellowish caterpillars. Soon after emergence the young caterpillars starts attacking the crops. They gradually become pink in colour hence, called "pink bollworm" and they bore into bolls and webs together two seeds to make a resting place especially for hibernation. This feature of two seeds attached together is known as ``double seeds" and in this condition the entire larval life is passed lasting for about 2-4 weeks. This is followed by pupation inside a flimsy silken cocoon in the soil or within the boll but outside the seed. The pupal period lasts about 10-14 days when the adult moth emerges. There are several generations in a year.

Control

The most effective method of physical control is the heat treatment in which the infested cotton seeds are heated to a temperature of 65^0C for 30 seconds. This will kill all the caterpillars inside the seeds.

Chemical control involves the use of endosulfan 35 EC @ 1.25-1.50 litres/ha in 1000 litres of water. Rest of the insecticides used are same as for spotted boll worm.

8. INDIGO CATERPILLAR

**Spodoptera exigua* Hubner (Lepidoptera: Noctuidae) (Fig. 49)

It is a polyphagous insect pest. It has been referred to as the indigo caterpillar because in earlier days when the indigo industry was flourishing, this pest used to be quite injurious to young indigo

* Same as *Laphygma exigua* Hub.

crop. The damage is caused by caterpillars which are gregarious, moving in swarms and destroy the young leaves of jute. It is also a minor pest of rice, cotton, tobacco, gram, maize etc.

Fig. 49. Indigo caterpillar - *Spodoptera exigua* **A.** Larva, **B.** Adult

Life history

The eggs are laid in clusters (up to 200 eggs) on the leaves. The egg clusters are covered with buff coloured hairs. The eggs hatch in 2-4 days. The larvae are blackish and there are six instars. Larval development takes 15-20 days. Pupation takes place in the soil and lasts for about 6 days. The adult is a small brown moth of wing span upto 25 mm. and they live for 8-10 days. The total life cycle is completed in about 21-30 days. There are several generations in a year.

Control

Cultural methods of control such as ploughing and burning the crop stubbles, flooding the infested field and the removal of weeds, all help to lower the pest population. The adult caterpillars can be controlled by spraying endosulfan 35EC @ 450-1500 ml/ha mixed with 500-1000 litres of water depending up on the severity, growth and type of caterpillars.

9. SUNNHEMP HAIRY CATERPILLER

Utetheisa pulchella Linn. (Lepidoptera: Arctiidae) (Fig. 50)

The caterpiller is known as "Sunnhemp hairy caterpillar". The caterpillar feeds on the leaves and capsules. The adult moth is creamy white with black and red spots on the fore wings and dark black spots on the margin of the hind wings. The full grown larva measures about 3.8 mm. in length and is provided with red, white markings on the body, the head is slightly brown in colour.

Fig. 50. Sunnhemp moth, *Utetheisa pulchella*

Life history

The female moths lay small whitish eggs on the tender leaves and shoots. Newly hatched larvae feed on the leaves. In the mature crop caterpillars feed on pods. Pupation takes place in the leaves as well as in the soil. The entire life cycle is completed in about 5 weeks. There are several generations in a year.

Control

The caterpillars can be controlled by spraying the crop with endosulfan 35 EC @ 1.5 litres/ha in 1000 litres of water. The caterpillars and moths can also be destroyed by hand picking.

10. JUTE SEMI-LOOPER

Anomis sabulifera Guen. (Lepidoptera: Noctuidae) (Fig. 51)

It is a specific pest of jute and is the most destructive in the jute tracts of India and neighbouring countries. It is considered to be the most serious pest of jute foliage. The infestation is said to come about in three successive waves of which the middle one is the most injurious. The apical buds of the plant are most vulnerable and the intensity of infection tends to vary inversely with the age of the foliage.

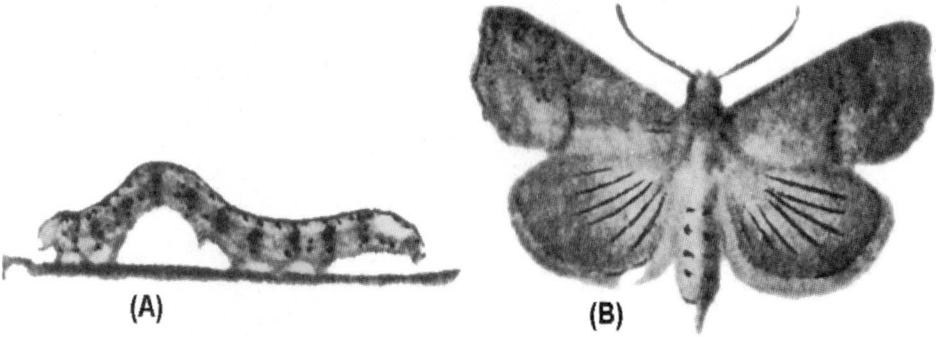

Fig. 51. Jute semilooper *Anomis sabulifera*, **A.** Larva, **B.** Adult

Life history

The pest hibernates during winter in pupal stage in the soil and the moths emerge with the first moist warm weather. The moths lay eggs on the jute plant singly particularly on the underside of the young leaves. The eggs look very much like a tiny drop of water on the leaf. There may be several such eggs on each leaf. A single moth can lay more than 150 eggs. The egg period lasts for about 2 days after which a small, green caterpillar hatches out. The tiny caterpillars on emerging start feeding on the apical leaves and the buds. The full grown larva is about 4 cm. and it attains this size after 5 moults during a period of about 2-3 weeks. The pupation may take place on the plant or in the soil. The pupal period lasts for only a week or so after which the moths emerge and begin to produce the next generation. The whole life cycle takes abut one month, thereafter several generations are completed in a year.

Control

The pest can be successfully controlled by spraying the crop with methyl – parathion 0.04%. Spraying endosulfan 35EC @ 450-1500 ml/ ha mixed with 500-1000 litres of water is also very effective.

11. JUTE STEM – GIRDLER

Nupserha bicolor postbrunnea **Dutt (Coleoptera: Lamiidae)**

This pest was reported to be serious on jute crops. The main damage is caused by the adult beetle while preparing sites for egg laying on the stem. It is distributed throughout the jute belt in India.

Life history

The female while preparing for laying eggs, cuts with the help of its sharp mandibles, two rings around the stem so that the distance between these two rings is generally 1-1.4 cm. and thereafter it makes, with its mandibles again a slit on the stem between these two rings. The depth of this slit reaches down to the pith level where a single egg is deposited. The average number of eggs laid per female is 35, hence the same number of jute stems are damaged within 2-3 weeks. The eggs are yellowish, about 1.5 mm. long and 0.5 mm. broad. They hatch in 3-4 days. On hatching, the larvae travel downwards along the central hallow area, resulting from the disintegration of the pith. The total larval period during the normal jute season varies between 30-50 days. The mature larva measures about 1.4 cm. and at the advent of winter the larva enters the diapause phase. Before pupation or diapause, the larva makes a pupal chamber within the hollow of the stem. The diapause phase lasts till a favourable combination of environmental factors is obtained during the following summer but under unfavourable conditions it may last for several years. The pupa closely resembles the adult beetle in form and size. It measures about 0.8 cm. The emergence of the adult largely depends on the time and amount of rain fall.

Control

The jute girdler can be controlled by spraying methyl demeton 25 EC @ 800 ml/ha or dimethoate (Rogor) 30 EC @ 600 ml/ha in 500 litres of water.

12. JUTE STEM WEEVIL

***Apion corchori* Marshall (Coleoptera : Apionidae) (Fig. 52)**

This stem borer is a weevil which is also commonly called the jute Apion. The weevil can cause appreciable damage to early sown jute or where the crop is grown for seed. It damages the quality of the jute fibre. The female weevil bores holes for oviposition by means of its rostrum in the jute stem.

Life hisotry

A single female has been observed to lay upto to 675 eggs during a period of 124 days. On an average an egg measures about 0.43 mm. in length and 0.33 mm. in breadth. The larvae emerge after 3-5 days and start feeding on the surrounding tissues. They are full-fed in 8-11 days. The full grown grub is about 2.85 mm. in length and about 1 mm. in breadth. The grub period may be as short as 8 days. The mature grub makes a rough chamber inside the stem and pupates therein. The pupa measures about 2.1 mm. in length and about 1 mm. in its maximum breadth. The pupal period may be 4-6 days after which the adult weevil emerges from the pupal case through an exit which is either made by the larva before pupation or by the weevil itself.

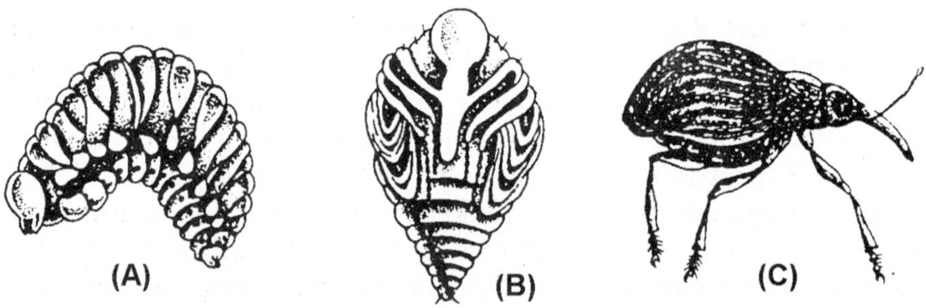

Fig. 52. Jute stem weevil, **A.** Larva, **B.** Pupa, **C.** Adult

The adult is a tiny weevil about 1.8 mm. in length and 0.8 mm. in breadth with a very conspicuous snout. It is dark brown or dull black in colour clothed with small whitish setae. The whole life cycle can be completed in about 15-24 days, but the maximum longevity of the adult has been reported to be as much as 209 days. Thus, there are a large number of over lapping generations during the jute season. Winter is generally passed in adult stage and the weevils concealing themselves in various bushes, shrubs and hedges and start laying eggs on the new crop next year.

Control

The pest can be controlled by spraying the crop with methyl parathion 0.03% which kills the grubs inside. The jute stem weevils both adults and caterpillars can be controlled by spraying endosulfan 35 EC @ 450-1500 ml/ha mixed with 500-1000 litres of water depending on the growth stage and type of caterpillars. The pest may be suppressed by removal and destruction of infested plants and by collecting and destroying the stubble after harvest.

16

PESTS OF SUGARCANE

1. SUGARCANE LEAF HOPPER

Pyrilla perpusilla **Walk. (Homoptera: Fulgoridae) (Fig. 53)**

This is a serious pest of sugarcane and it prefers the board leaved succulent variety as the food plant. Both the adults and nymphs are voracious feeders which ingest the sap from the underside of leaves. This results in the shrivelling and drying up of the leaves. A serious attack affects the quality and yield of the crop and the sucrose content is reduced to 50% and Gur becomes tasteless.

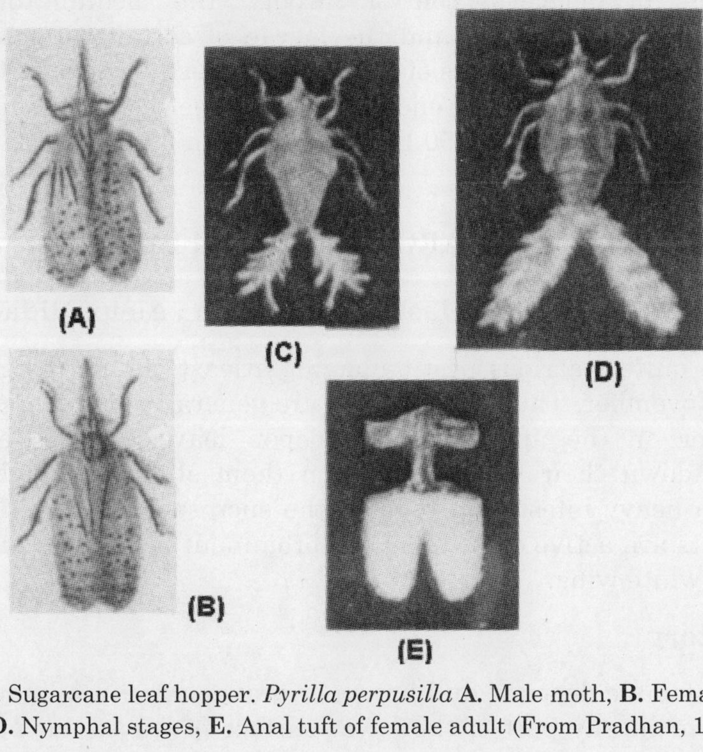

Fig. 53. Sugarcane leaf hopper. *Pyrilla perpusilla* **A.** Male moth, **B.** Female moth, **C-D.** Nymphal stages, **E.** Anal tuft of female adult (From Pradhan, 1969)

The adult hopper is normally dull yellow in colour with a head produced anteriorly into an upwardly turned snout. The two pairs of wings are folded and rest over the back, a characteristic feature of Homoptera.

Life history

The female lays greenish yellow eggs about 300-530 in clusters on the lower surface of leaves or in leaf-sheaths or on stem. The number of eggs in each cluster vary from 20-60. All these eggs are covered over with white cottony waxy filamentous threads as a safeguard from parasite attack. Hatching occurs in about 7-29 days depending on temperature and pale brown nymphs are released. These are covered on their anal processes with waxy filaments. The nymphs undergo five moults after which fully developed adult individuals are formed. The entire life cycle takes about 46-59 days. In a year, the first brood crops up in April to June, the 2nd in July to September and the 3rd in September to November.

Control

Occasionally mechanical control by collection and destruction of eggs between fingers on the leaves or by burning is applied. Adoptation of biological control through the medium of certain Hymenopterious parasites and the larvae of certain coccinellides is another successful measures of lowering the pests incidents. Chemical control involves the use of endosulfan or dimethoate or fenitrothion 0.025% spray or malathion 50 EC or endosulfan 35 EC @ 1.25 litres/ha in 1000 litres of water.

2. SUGARCANE WHITE FLY

***Aleurolobus barodensis* Mask. (Homoptera : Aleurodidae)**

It is quite a serious pest damaging atleast 50% of the crop from July to November. The insects which are generally found on the lower surface or on the apices of the unopen leaves, desap the plant, bringing down their vigour and turn them sticky and pale. As in *Pyrilla*, a heavy infestation reduces the sucrose content of the plant. The adults are active creatures with prominent black eyes and a pair of mealy white wing.

Life history

The female deposits dull brown or orange eggs in a line near the mid ribs or in any part of the lower surface of leaves during winter.

The number of eggs laid vary from 60-65; the eggs have a broad pedunculated base and a pointed tip. They are found in groups of 15-20. They hatch in about a week or so into dull yellow nymphs. In course of time the colour of nymphs changes to shiny black. The locomotory legs of the nymphs which are functional for a short duration atrophy immediately after the first moulting, and the insect fixes itself to some point on a leaf and remains there till it completes its life history. There are usually four instars after which the nymph enters into a "pupal" stage, an unusual features in the metamorphosis of Hemiptera-Homoptera lasting about 10-15 days. Finally the pupa transforms into the adult whitefly which lives 24-48 hours only. The whole life cycle occupies 25-55 days. This pest is very active from August onwards. There are several generations in a year.

Control

Both mechanical and biological controls are commonly employed. The pest can be controlled by spraying the crop with malathion 0.1% or endosulfan 35 EC or monocrotophos 40 EC @ 1.5 litres/ha mixed with 800 litres of water.

3. SUGARCANE TOP SHOOT BORER

Scirpophaga nivella **Fabr. (Lepidoptera : Pyralididae) (Fig. 54)**

It is a major pest of sugarcane attacking the crop at all stages of growth and causing a heavy damage in July and August. The larva usually attacks the top-shoots of the cane and hence is known as 'top shoot borer'. This results in the curling and drying up of leaves and formation of spiral structures known as deadhearts. The adult insect is a silvery white moth with dark black eyes and tufts of yellow, orange, brown or red hairs over the anal segment in females.

Life history

The female lays 60-80 eggs, oval or elongate in shape, in 5 – 8 clusters on the underside of leaves. Hatching occurs in about 6-12 days and very small shiny caterpillars are released. Immediately after emergence, these caterpillars start attacking the rolled up leaves. Gradually they bore into the mid rib of an upper unfold leaf and enter the terminal shoot through the rolled base of the top leaves. The entire larval period in this manner is passed either on leaves or on top shoots and in about 22 days the larva becomes full grown possessing a creamy white colour with dark streaks and a few scattered hairs all over the body. It is in this larval stage characteristic tunnels are made inside the terminal shoots.

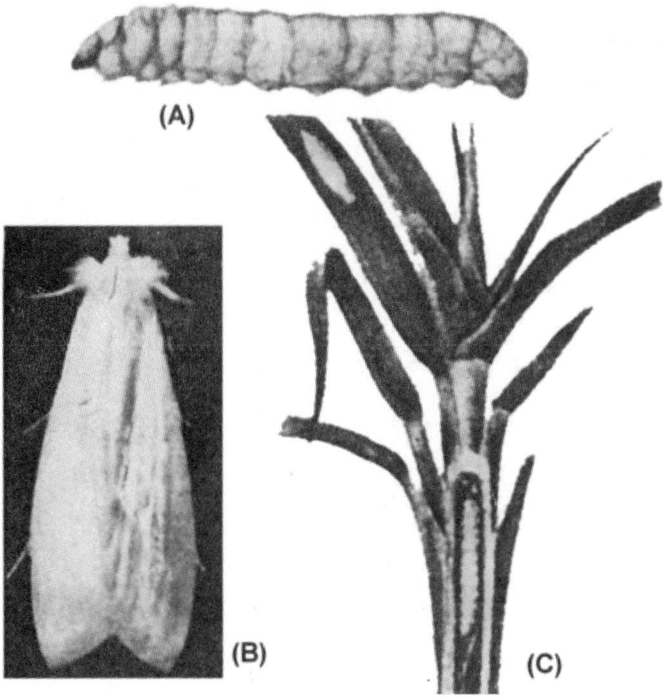

Fig. 54. Sugarcane top shoot borer *Scirpophaga nivella* - **A.** Larva, **B.** Adult, **C.** Attacked shoot of sugarcane showing larva inside; note dead-heart and egg-mass on leaf (From Pradhan, 1969)

A full grown larva cuts off portions of the plant at right angle to the original tunnel and makes necessary arrangements for the ensuring pupa to rest. The pupal shelter is chiefly constructed by the caterpillars by chipping up a small circular portion from the bark of the cane. This is gummed in its place by the secretion of the mouth, thus forming a close exist hole. The cavity thus formed is lined with silken threads secreted by larva. It is in this cavity the larva changes into a pupa and remains in this condition for 7-11 days. At the end of this period the adult moth emerges. The whole life cycle takes about 40-50 days but varies when conditions become unfavourable. The pest completes 4 to 5 genrations in a year.

Control

As in the case of other pests, both mechanical and biological controls are employed. Biological control is brought about by the introduction of Hymenopteran parasites like a Braconid wasp or a Chalcid wasp e.g. *Trichogramma minutum*. Spraying the crop with 1.5

litres of endosulfan 35 EC or monocrotophos 40 EC mixed in 1000 litres of water on leaf whorl from the top is very effective.

4. SUGARCANE ROOT BORER

***Emmalocera depressella* Swinh. (Lepidoptera : Pyralididae) (Fig. 55)**

The greatest amount of damage is caused by the caterpillars. The attack is confined to the base of the newly formed stem. Presence of the dead hearts or dry central shoot among the young cane is the first symptom of this pest. The damage done by root borer is very characteristic because the caterpillar enters at the base of the stem, close to the surface soil. These borers are confined to base of the shoot which is bored across side ways or round about. On pulling the whole of the affected shoot or even the entire cane breaks exposing the caterpillars. In grown up canes, they may bore the thick roots but seldom attack the base. Damage is serious from the end of April to the end of September or middle of October. The young canes die and the older ones dry up and fall down. Sugar contents is greatly reduced. The moth is cryptically coloured hence cannot be easily detected even in worst infested fields. The female moth measures about 30-35 mm. along wing expanse. The hinder wings are shorter and wider having a dark longitudinal stripe.

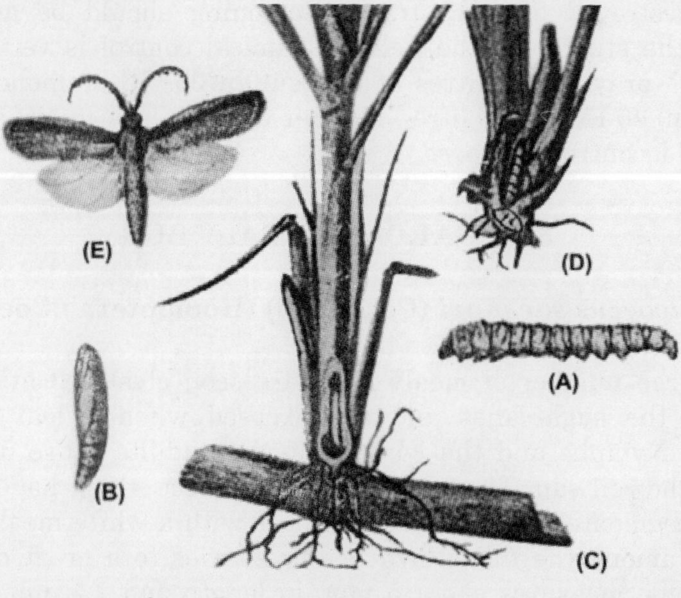

Fig. 55. Sugarcane root borer *Emmalocera depressella* - **A.** Larva, **B.** Pupa, **C.** Cane shoot cut open to show attack by larva, **D.** Cane plant cut open to show pupa. **E.** Adult (From Pradhan, 1969).

Life history

Creamy white, scale like flattened eggs are laid singly or in groups on leaves and stems in the first week of May. A female lays as many as 277-355 eggs. Young larvae hatch out in about 4-8 days which are 2 mm. long with yellow body and yellowish brown head. They bore the base of the young seedlings, getting entrance through sides near or below the surface of the soil. They are confined to the new stems where they grow, and fully mature larva measures about 30 mm. in length with a creamy white wrinkled body with brown head. The larval duration is about 30-50 days. The borer when left underground, attacks other side shoots and plants. The caterpillar is active in cane from the time, the cane is three weeks to a month old. Before pupation the caterpillars come up above the ground level in the stem, makes an exit hole in the form of silken tube which reaches the surface of the soil. The pupae are reddish brown in colour. Pupal stage lasts about 9-14 days. They hibernate in winter months in the larval stage in roots. The complete life cycle takes about 56-69 days. There are about 4 generations in a year.

Control

Prompt removal of dead hearts is quite effective and the moth can be destroyed by light traps. Ratooning should be avoided by burning the stubble in the fields. Chemical control is very difficult. However, spraying 1.5 litres of endosulfan 35 EC or monocrotophos (Nuvacron) 40 EC in 600 litres of water when eggs or moths are visible in the field, controls the pest.

5. SUGARCANE MEALY BUG

Saccharicoccus sacchari (Cockerell) (Homoptera : Coccidae)

A large number of mealy bugs are seen clustered at the basal nodes of the sugarcanes and are exposed when a leaf sheath is removed. Nymphs and the wingless female adults cause damage by sucking the cell sap. They are inert pink insect which have a rotund sac like segmented body covered all over with a white mealy powder. The size among the full grown females varies to a great deal but a large female measures about 5 mm. in length and 2.5 mm. in width. The male are sluggish, have one pair of wings and short lived and harmless.

Life history

The pest breeds throughout the year and lays a large number of eggs at short intervals. The eggs are very small yellowish smooth, cylindrical and rounded at both the ends and measure 0.35 mm. in length and 0.16 mm in width. Within a few hours the crawlers emerge. The tiny young ones are pinkish in colour and very active. On finding a suitable host plant they force themselves below the leaf sheaths near the basal node. As the canes grow taller, the older bugs remain at the lower end and the crawlers reach the higher nodes. Nymphs feed voraciously and pass through 6 stages before they are full grown in 2-3 weeks. The life of a full grown female, is approximately 3-5 days. The entire life cycle is completed in about a month during summer. The pest completes several generation in a year.

Control

The pests can be controlled by spraying the crop with malathion 0.05% or diazinon 0.03%.

6. SUGARCANE BLACK BUG

***Cavelerius excavatus* (Dist.) (Heteroptera : Lygaeidae)**

The black bug is distributed throughout India. In addition to sugar cane, as with many pests, it has also been recorded feeding on rice, maize, wheat, joar and grasses. The damage is caused by both the adults as well as nymphs which suck the cell sap. The leaves become chlorotic, bearing brown patches. The tips and the margins of the leaves start drying up gradually. In severe infestation the whole plant may wither. Further, as a result of heavy infestation the quality of juice is also severely affected.

Life history

The insect breeds throughout the year but very common in winter. In summer, each bug lays 55-478 eggs in groups or clusters on the inner side of the leaf sheaths of the sugarcane. Each cluster may contain 14-67 or more eggs. The eggs are creamy white in colour which hatch in about 9-11 days during summer and in about 3-5 months during the winter. The nymphs passes through 5 instars and complete

* Same as *Macropes excavatus* Dist.

their development in about 4-10 weeks depending on the climatic conditions. The adult bugs are black in colour and about 6-7 mm. long. The wings are provided with white patches which extend beyond the abdomen. The insects complete about 3 generations in a year. Adults have a long life. Males may live for 4-7 months while the females live for 4-8 months. In winter eggs are laid in the soil about 5-7 cm. below the surface near the plant base which hatch during next spring. Thus in the spring season, the new nymphs and old adults are seen feeding side by side.

Control

The pest can be controlled by spraying the crop with malathion 0.05%. Spraying should be directed from the top to down, towards the central whorl and the leaf sheath. Similarly, spraying endosulfan 35 EC or monocrotophos 40 EC @ 2 litres/ha mixed in 800-1000 litres of water is effective to control the pest. Furthermore, burning of the stubbles has also been found to be very useful in controlling the pest population.

17

PESTS OF OILSEED CROPS

1. CASTOR SEMILOOPER

******Achaea janata* **Linn. (Lepidoptera : Noctuidae) (Fig. 56)**

This is one of the most destructive pests of castor crop. Damage is caused by the caterpillars which feed voraciously on the delicate parts of the leaves and at the end only the veins remain. Due to considerable loss of leaves, the castor seed yield is also greatly reduced. The peculiarity of *Achaea janata* is that in the larval stage, it is a serious pest of castor and in adult stage it is a fruit sucking moth and is a serious pest of *Citrus* fruits, a rare feature among moths, which are normally harmless.

Fig. 56. Castor semi-looper *Achaea janata*, **A-C.** Adult forms, **D.** Semi looper (From Pradhan, 1969)

---

* Also written as *Achoea janata* Linn.

The adult of *Achaea janata* is a pale reddish-brown moth with a wing expanse 60-70 mm. The wings, both the fore and the hind are decorated with broad zigzag markings and a large yellowish area and dark brown patches. The mature larva is dark and is marked with blue black, yellow and reddish stripes and a dorsal hump near the head end of the body. There is a characteristic white mark on the head. The colour pattern is very different in this insect and some workers recognize four different varieties of caterpillars in the 3rd and 4th instars. They are very distinct against the background of the plants.

Life history

Each female moth can lay several hundred eggs. The total number of eggs laid by a female varies from 450-500 which are deposited singly on the leaves. The egg is about 1 mm. in length and has numerous ridges and furrows on the surface, radiating from a circular depression at the apex. On account of its cryptic blue-green colour, it is extremely difficult to spot out the eggs on the leaves. Hatching of the larva takes place within 2 to 5 days. The first instar larva is extremely small and measures about 3.5 mm. in length. The larva emerges by biting a hole into the egg shell; the latter also forms the first meal of the newly hatched larva before it begins to feed on the leaves. The larva feeds and moults 4 or 5 times and become full grown in about 15-25 days. The mature larva measures about 60 to 65 mm. in length. The larva after searching a suitable hiding place prepares a loosely knitted cocoon of silk. Occasionally soil particles also get attached with the cocoons. The pupal duration is about 10-15 days and the moth, on emergence feeds on soft fruits of citrus, mango etc. There may be several generations in a year.

Control

Semiloopers can be controlled by spraying Thiodan (endosulfan) 35 EC @ 1.25 litres in 800 litres of water/ha. The use of insecticides is not recommended in the case of castor crop meant for the rearing of the erisilkworms. The semiloopers should be regularly collected and destroyed.

2. CASTOR HAIRY CATERPILLAR

Euproctis lunata Walk. (Lepidoptera : Noctuidae)

The castor hairy caterpillar is distributed throughout India. The caterpillar feeds on the leaves of various host plants and in the case of severe infestation they may cause complete defoliation. The attacked plants remain dwarf (under developed) and produce very little seed. Thus it is the larvae which cause damage to the castor plants.

Life history

The eggs are laid in the late winter i.e. February in clusters on the under side of the leaves. The eggs are covered with brown hairs from the anal tuft of the female. They hatch in 5-7 days. The young larvae feed gregariously for about a week and then disperse and feed individually. They pass through 6 stages and are full-fed in 2-3 weeks. The mature larvae measure 35-40 mm. dark grey in colour with a white stripe on the dorsal side. They have long hair all over the body. They make loose silken cocoons in the plant debris lying on the ground and pupate inside. The pupal stage lasts about a week in summer. During the winter months the egg, larval and pupal stages may be prolonged. The moths are pale yellow and are seen sitting on castor and other neighbouring plants during day-time and are harmless.

Control

The young caterpillars can be killed by applying 5% malathion or carbaryl dust to the crop. The grown up caterpillars can be killed with methyl parathion 0.04% sprays.

3. CASTOR CAPSULE BORER

***Dichocrocis punctiferalis* Guen. (Lepidoptera : Pyralididae)**

The castor capsule borer is distributed throughout the country. The damage is caused by the caterpillars which bore into the capsules. The mature caterpillar measures 25-30 mm. in length and is reddish-brown in colour with black blotches all over the body and a pale stripe on the lateral side, the moths are orange yellow with black markings on both the wings. Since the capsules and the seeds are damaged from inside, the yield is greatly reduced.

Life History

The moths lay eggs on the leaves and other soft parts of the plant. The eggs hatch in about a week. The larvae pass through 4-5 instars and are full fed in 2-3 weeks. Pupation takes place inside the seed or in the frass that collects after feeding. The pupal stage lasts about one week. The life cycle is completed in 4-5 weeks. Three generations are completed in one year. This pest is active on caster from September to March.

Control

The castor capsule borer can be controlled with the repeated applications of carbaryl 0.2%. In the capsule bearing crop application of insecticides is very difficult. It has been advised to collect and destroy the infested shoots and capsules.

4. MUSTARD SAWFLY

Athalia proxima Klug (Hymenoptera : Tenthredinidae) (Fig. 57)

It feeds on various cruciferous plants like mustard, cabbage, cauliflower, turnip, radish etc. Damange is done by the larvae which are dark green, measuring about 16-18 mm. in length. It is a very harmful pest of mustard and other crops of the family Cruciferae. The pest is distributed practically throughout the India. The activity is confined from October to March. It is in the larval stage that the pest damages the crop by voraciously feeding on the leaves of the young crop. The adult stage is a rather thick-set insect with a mixture of yellow-brown markings on the body and a dark reddish-brown colour on the wings.

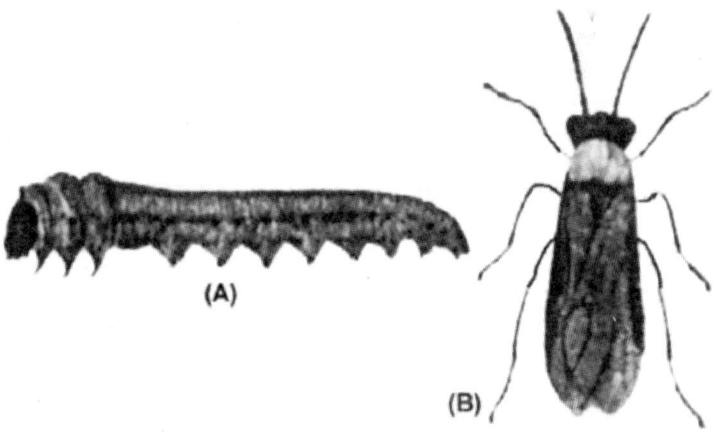

Fig. 57. Mustard sawfly *Athalia proxima* - **A.** Larva, **B.** Adult (Lefroy, 1909)

Life history

The female lays 30-35 eggs singly in slits made with the saw like ovipositor along the leaf margins on the underside. The eggs hatch in about 4-8 days and the larvae feed exposed in groups of 3-6 on the leaves during the morning and the evening. They pass through seven stages and are fully grown in 16-35 days. The full grown larva measures 16-18 mm. in length which enters the soil and pupate there in a water proof, oval silken cocoon, and emerges as adult in about 11-37 days. Thus the life cycle is completed in about 31-74 days.

Control

Spray the crop with metasystox 25 EC @1 litre/ha or dimethoate (Rogor) @1 litre/ha in 1000 litres of water or malathion 50 EC or thiodan 35 EC @1-1.25 litres/ha in 1000 litres of water.

5. MUSTARD APHID

******Lipaphis erysimi*** Kalt. (Homoptera: Aphididae) (Fig. 58)**

The most serious pest of the mustard crop is the aphid, *L. erysimi.* This species has a wide distribution in the world. Besides brassicas to which the mustard belongs, this pest attacks a number of other economic plants, particularly to those of the family Cruciferae. The attack commences on plants even when they are in seed-beds as well as at all subsequent stages of their growth. They suck the sap of the plant by thrusting their needle like stylets into the tissues. Such a damage renders the attacked crop pale, weak, exhausted and incapable of bearing any seed. Leaves are most susceptible, those which are affected curl, shrivel and form cups, and finally drop. The insect pours a good deal of sugary excreta known as honey dew which makes the infected plant sticky to touch and facilitate the growth of a black fungus on them. This gives the appearance of dirty black leaves with shrivelled up lamina on the plant.

The adult insects are very small having light green or slightly yellowish colour. They have a soft body and are of oblong in shape. One of the special morphological features in these insects is the presence of a pair of short tubes known as honey tubes or cornicles which project from the upper surface of the hind end of the body. It is through these tubes that honey dew is excreted out.

Life history

This insect is most abundant in winter i.e. from December to March. In the summer aphids survive on some other cruciferous plants of minor importance. The aphid breeds parthenogenetically and the females give birth to 26-133 nymphs. They grow very fast and mature in 7-10 days. Thereafter, they begin to give birth to another

* Same as *Rhopalosiphum erysimi* (Kalt.)

generation. There are about 45 generations in one year. Cloudy and cold weather is very favourable for the multiplication of this pest. Generally, apterous forms are present but during migration of the insect species in spring, winged forms are produced.

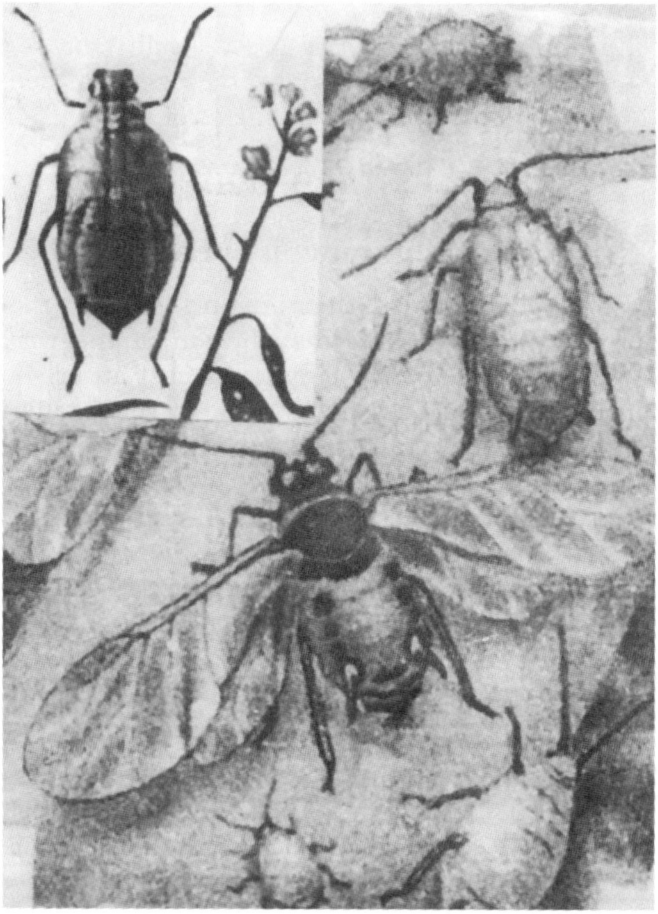

Fig. 58. Aphids (Inset mustard aphid *Lipaphis erysimi*)

Control

Mustard aphid can be controlled by spraying the crop with any one of these insecticides: nicotine sulphate 0.05% diazinon/methylparathion/phosphamidon 0.025%, dimethoate 0.03%, methyl demeton or malathion 0.1% or 0.04% monocrotophos 40 EC. Basal application of thimet 10% granules @ 10 kg/ha is also useful. In the flowering season, spraying should be carried out in the evening, when most of the pollinators including honey bees stop visiting the crop. When the crop is in the flowering stage, phosphamidon and dimethoate which are comparatively safer to pollinators, should be preferred.

6. PAINTED BUG

***Bagrada cruciferarum* Kirk. (Heteroptera:Pentatomidae) (Fig. 59A)**

It is a serious pest of many of the cruciferous plants as well as those of the cucurbitaceous species, the seriousness being sporadic. All the stages in the insect's life cycle cause damage to the plants. The individuals suck the sap from the leaves which consequently turn yellow and drop off exposing the plant to bacterial and fungal infection.

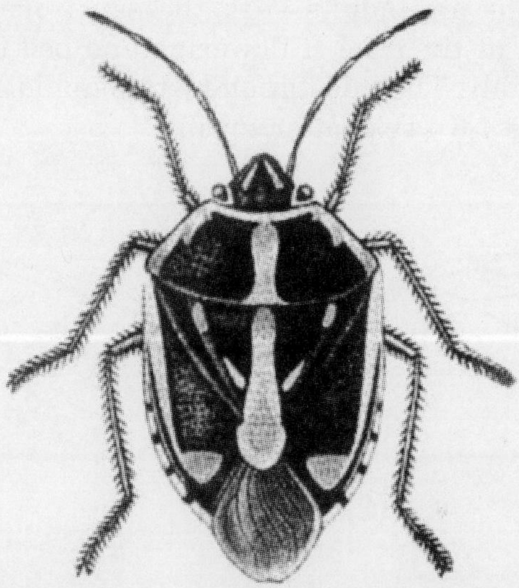

Fig. 59A. Painted bug, *Bagrada cruciferarum*

The adult is black in colour painted with irregular red markings. The scutellum is very large. The adults are extremely tenaceous and can withstand the absence of food for a fairly long period.

Life history

The female lays about 230 eggs singly or in groups of 2-12 on the leaves and stems. The eggs which gradually become red in colour hatch in about a weeks time into young greenish red nymphs. They immediately start feeding on the plant and gradually change colour. At the same time they moult also, the number of moults varying often. The nymph transform into adult in about 3 weeks time. The activity of pest ranges from October to March with 6-7 broods in a year.

---

* Same as *Bagrada picta* Fabr.

Control

The pest can be controlled by spraying the crop with any of the insecticides: viz. diazinon 0.02% or malathion 0.05% or dimethoate 0.03%.

7. GROUND NUT APHID

Aphis craccivora Koch. (Homoptera: Aphididae) (Fig. 59B)

It is a serious pest of ground nut and is distributed throughout India. Both nymphs and adults suck the sap from the underside of leaves. The attack at the time of flowering and pod formation reduces the yield considerably. The infestation by this aphid on the ground nut crop usually occurs 4-6 weeks after sowing.

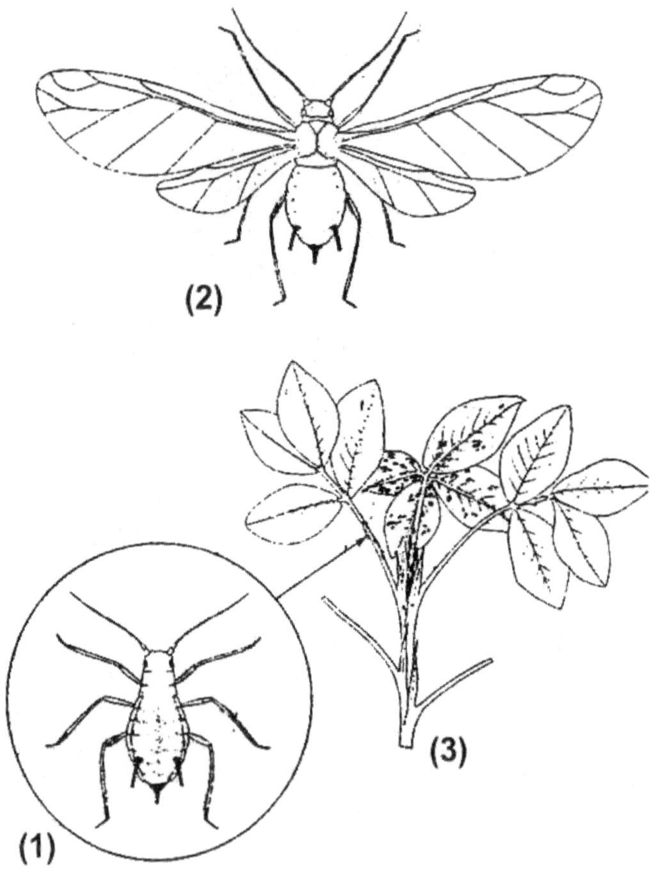

Fig. 59B. Ground nut aphid, *Aphis craccivora*.
1. Nymph, **2.** Adult, **3.** Infested plant (From Hill, 1975)

Life history

The offsprings of the winged forms may be wingless or even parthenogenetic. The female may produce 8 to 30 young ones in a life span of 10-12 days. The young nymphs are brownish in colour and pass through 4 moults to become adult in about 5-8 days. The wingless females start producing the brood within 24 hours of attaining that stage. Breeding occurs – throughout the year and both forms, winged and wingless are present.

Control

The pest can be controlled by repeatedly spraying the crop with any of the insecticides viz. malathion 0.05%, carbaryl 0.1%, endosulfan 0.05%, dimethoate 0.05%, or methyl demeton 0.05%.

8. GROUNDNUT LEAF MINER

***Stomopteryx nerteria* Meyr. (Lepidoptera: Gelechiidae)**

This pest is fairly specific to ground nut, although it has been reported infesting a number of other plants also in different parts of the country. The damage is caused by the small caterpillars which mine into the leaves, skeletonize them and web them together.

Life history

The adult stage is a small moth rather bronzy in colour and with a wing span of less than 1 cm. Its activities are nocturnal, it is strongly attracted to light, during the day it remains hiding under clods or in crevices. It lays eggs singly on the under side of the leaves of its food plants and one female can lay up to several hundred eggs. The eggs hatch after about 3 days. The newly hatched larva is about 1.5 mm. It wanders about for a little while and then it mines into the leaves and later on bites its way out. It then webs together a number of leaflets and feed inside the chamber so formed. The larvae are full grown in 9-17 days. The full grown larva is 6 to 8 mm. in length and dirty-green in colour. Pupation takes place in silken cocoon, which is about 9 mm. in length. Pupal period is about 4 days. Under suitable conditions (food, climate etc.), this insect can breed throughout the year.

Control

It can be controlled by spraying the crop with carbaryl 0.2%. The pest can also be controlled by spraying the crop with metasystox 25

EC or dimethoate (Rogor) 30 EC @ 1 litre/ha in 1000 litres of water or 250 ml/ha of dimecron in 1000 litres of water.

9. TIL HAWK MOTH

Acherontia styx West. (Lepidoptera: Sphingidae) (Fig. 60)

This insect is commonly referred as "Hawk Moth" "Sphinx Moth" or "Death's Head Moth". The larvae feed on the leaves and defoliate the Til plants (*Sesamum indicum*). Besides Til, it has also been found to damage potato, brinjal, lab-lab and the ornamental plants like balsam. The moth is also harmful to apiarists because it sucks honey from the honey-combs, although at times it is stung to death by the bees. The adult of *Acherontia styx* is a large reddish – brown moth, with a wing span of at times more than 10 cm. The forewings possess a mixture of dark- brown and grey ornamentation with dark blackish- wavy markings and a prominent yellow spot on each wing. The hind wings have an ochre background on which there are two broad dark brown wavy cross stripes. The head and thorax are also dark brown and on the latter there is a prominent Death's Head mark. The abdomen is ochracious in general hue with drak-brown cross stripes.

Life history

The moths lay eggs singly on the under surface of the leaves of the food plant and the eggs are fairly large, their colour is at first greenish white but later on it turns yellow. Hatching takes place within a few days and the just hatched larva is pale- yellowish in colour. Soon it begins to feed on the leaves and develops broad green stripes running obliquely along the dorso-lateral surface. The full grown caterpillars which attain a length upto 5 cm. or more and also a thickness of more than 1 cm. are in the habit of retracting and drawing in the head and some of the anterior segmants of the body into a sort of Sphinx- like posture, this along with a horn-like projection from the hinder end of the abdomen enhances the monstrosity of these larvae which otherwise have quite a plump body with a pleasent mixture of soft colours. The larval period is un-usually long and at times it lasts 2 months or more. The pupa is also almost equally long lived. The winter is passed in the pupal stage in the soil. There are 3 generation in a year.

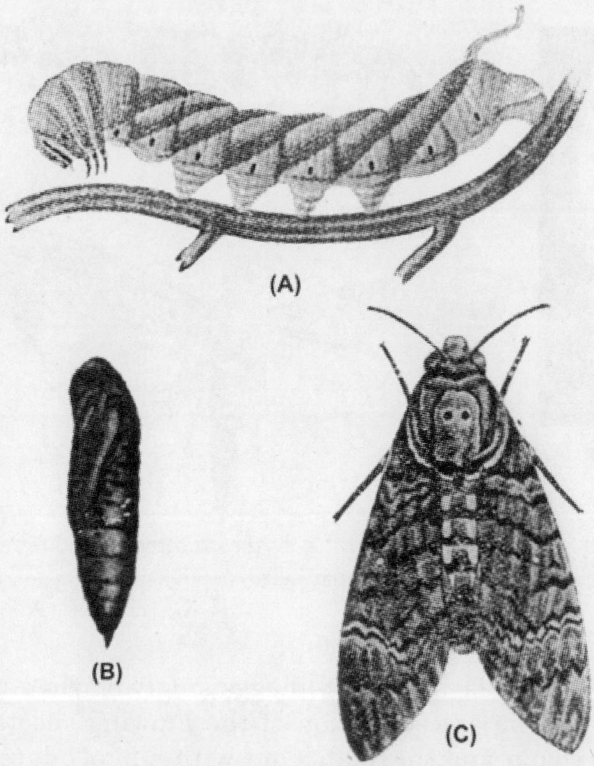

Fig. 60. Til - Hawk moth *Acherontia styx* - **A.** Caterpillar, **B.** Pupa, **C.** Moth (From Lefroy, 1909)

Control

The Hawk moth larvae can be controlled by spraying the crop with endosulfan @ 1.25 litres/ha in 1000 litres of water or quinalphos 25EC @ 500 ml/ha in 1000 litres of water. Hand picking is also a good method to suppress the larvae.

10. TIL LEAF AND POD CATERPILLER

Antigastra catalaunalis Dupon. (Lepidoptera: Pyralididae) (Fig. 61)

This is a serious pest of Til (*Sesamum indicum*). The larval stage damages both the young apical shoot and the young pods. The adult stage is a medium sized moth with a wing span of 15-20 cm. The fore wings are reddish yellow in colour with zigzag indistinct reddish decorative markings on them, the hind wings are pale-yellow and transparent. They are nocturnal in habit.

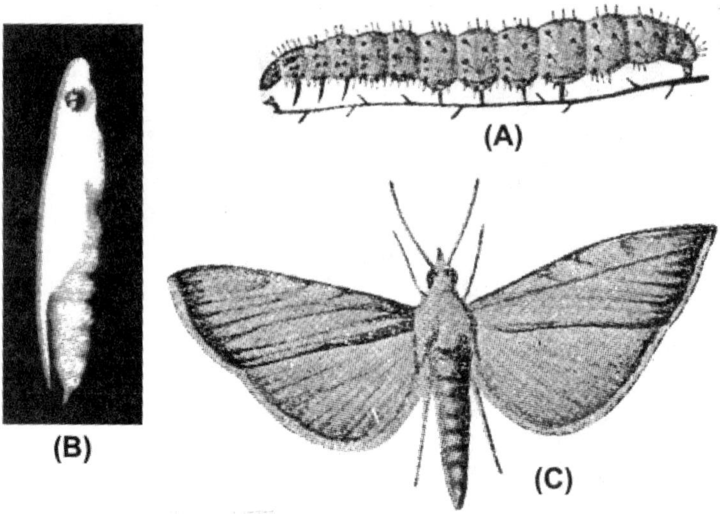

Fig. 61. Til leaf and pod caterpillar *Antigastra catalaunalis* **A.** Larva, **B.** Pupa, **C.** Adult (From Lefroy, 1909)

Life history

The female moth lays about 140 eggs singly on the under side of the leaves and on the apical portion of the growing shoots. The eggs are greenish in colour and they hatch out within 2 to 7 days depending on the season. The just hatched larva is about 2 mm. in length and somewhat whitish in colour. Later on, the colour changes to green and the mature larva is about 15 mm. in length, is also provided with dark brown markings. Larval period lasts from 10 days to more than a month. Pupation takes place in cracks, crevices or under the fallen leaves. The just formed pupa is green but becomes brown later on. Pupal period varies from 4 to 20 days. Thereafter the moth emerges which begins to lay eggs for the next generation within 4-5 days after emergence. The pest completes about 14 generations in a year depending on the temperature and humidity.

Control

The pest can be controlled by spraying the infested crop with carbaryl 0.1% against the moth to prevent them egg laying. Mechanical and physical methods can be also used. For controlling almost all the insect pests of oil seed crops like aphids, leaf miners, thrips and bugs, spraying the crop either with methyl demeton 25 EC @ 1 ml/litre of water or dimecron 100 EC @ 1 ml/ha in 1000 litres of water or dimethoate 30EC @ 1 litre/ha in 1000 litres of water is recommended. Spraying should be repeated if needed.

18

PESTS OF PULSE CROPS

1. GREASY CUTWORM

[*]*Agrotis ypsilon* Rott. (Lepidoptera: Noctuidae) (Fig. 62)

It is commonly known as ``greasy cutworm'', and is cosmopolitan in distribution. The attack is extremely serious during winter when young seedlings, roots and leaves are nibbled at night and the stems cut at the level of the ground by the caterpillars. Besides gram, it is also a pest of potato, tobacco, peas, cotton, tomato, vegetables and other winter crops. The caterpillars are polyphagous. The moth is dark brown with a grey body, grey fore wings with dark brownish black markings. The hind wings are almost white basally but with a dark terminal fringe particulary in the males. They measure about 40-50 mm. along the wings. The moth is nocturnal.

Fig. 62. Greasy cutworm - *Agrotis ypsilon* **A.** Larva, **B.** Adult

Life history

The female moth lays eggs in clusters, consisting of about 30 eggs in each underside the leaves. Several thousand eggs can be laid by a single female moth. The eggs are white, globular and ribbed, 0.5 mm. in diameter. The larva hatches in about 2-9 days and begins to feed on

[*] Also written as *Agrotis ipsilon* Hufn.

egg shell which forms the first meal of the larva. The larvae are also nocturnal and they remain underground in day time. The full grown caterpillar measures about 35-45 mm. long, is dull grey or brownish in colour. Larval development takes 28-34 days. The pupae are dark brown, 20 mm. long. Pupation takes 10 to 30 days or more depending on the temperature and then moth emerges and starts new life cycle. The life cycle from egg to adult takes about 32 days at 30°C and 67 days at 20°C.

Control

Removal of weeds from the field helps to eliminate oviposition sites. Fenitrothion has generally given best control of many cut worm species.

The pest can be controlled by application of endosulfan 5% dust @ 25 kg/ha or thiodan 0.05% spray.

2. GRAM CUTWORM

***Agrotis flammatra* Schiff. (Lepidoptera : Noctuidae) (Fig. 63)**

The gram cutworm is one of the most important pests of gram in India. It is a polyphagous pest and besides gram, it feeds on a wide variety of plants belonging to different families. The adult moth is a stout, greyish brown in colour measuring about 5-6 cm. in wing expanse. The fore wings of the moth have characteristic markings and smoky patches, the proximal areas being pale or yellowish in colour. Each wing is provided with a semi-circular spot below the pale or yellowish area and a greyish-brown kidney shaped spot towards the apical area. The damage is caused only by the caterpillars.

Life history

The moths lay yellowish white eggs on the under side of the leaves, on shoots, stems or in the soil within 5-10 days of their life which is about 7-13 days only. A female moth lays about 950 or more eggs in her life time. Hatching takes place in about 4-14 days depending on the temperature. The larvae are full grown in about 4-7 weeks and measure 40-50 mm. in length, dark grey or light green in colour. They pupate in the earthen cells constructed by them in the soil. The pupal stage lasts about 10-15 days in favourable weather but during winter it extends upto 5 weeks or more and after which the moth emerges. The whole life cycle is completed in about 6-13 weeks, depending on the climatic conditions. There are usually 2 generations in a year.

Fig. 63. Gram cutworm, *Agrotis flammatra* Moth.

Control

The pest can be controlled with lindane dust 2%. Cutworms can also be controlled by spraying of 35 EC endosulfan in the concentration of 2 ml/litres of water at 15 days intervals. Two or three sprayings are sufficient.

3. GRAM POD BORER

***Heliothis armigera* (Hubner) (Lepidoptera: Noctuidae) (Fig. 64)**

The "gram pod borer" is universally distributed and is a polyphagous pest. It is one of the most serious pests of gram. The insect has also been found damaging many other crops like tomato, ground nut, tobacco etc. In U.S.A. and other countries it is a serious pest of corn and is known as "cornearworm" or cotton and is called cotton bollworm. The adult is a stout built Noctuid moth with a wing span of 30-40 mm. The colour of moth varies from olive grey to red brown. The forewing has a dark speck and a dark area near the outer margin. The hind wing has a distinct smoky dark outer border. The damage is caused by the caterpillars.

Life history

Eggs are laid singly on the tender parts of its food plants. The eggs are spherical, sculptured and with greenish shine when laid and measure about 0.5 mm. in diameter, as the embryo develop the eggs turn brown. Depending on temperature the egg period lasts from a few

* Same as *Helicoverpa armigera* (Hub.)

days to more than a week. Each female may lay about 1000 or more eggs in her life time.

Fig. 64. Gram pod borer *Heliothis armigera* - **A.** Caterpillars on gram plant eating into pods. **B.** Pupa in the underground cell. **C.** Moth (From Pradhan, 1969)

The young larvae begin to feed on the tender portions of the leaves and shoots. Later, the caterpillars feed on the developing seeds after cutting hole in the pod with head inside but with the posterior part of the body outside the pod. They also burrow into the large pod and eat developing seeds. Their colour and size vary considerably. For example, the colour pattern consists of variable combinations of pale-green and pale-brown with longitudinal stripes along the sides. The full grown larva measures about 35-40 mm. in length. The larval period lasts about 2-3 weeks. Full grown larva burrows into the soil and pupates there in an earthen cocoon. The pupa is shining dark brown in colour measuring about 12-16 mm. in length. The moth emerges in about 10-14 days after pupation. Egg laying starts about 4 days after emergence and may continue for about 9 weeks or more. There may be as many as eight generations in a year.

Control

It can be controlled effectively by spraying the crop with 0.1% carbaryl or 1.5 litres of endosulfan 35 EC or 750 ml monocrotophos 40 EC/ha in 1000 litres of water. The caterpillars can also be collected and destroyed by simple hand picking. Crop rotation is also extremely useful.

4. RED GRAM POD FLY

******Agromyza obtusa*** Mallas (Diptera : Agromyzidae) (Fig. 65)**

It is a small metallic black fly which is also known as Tur Pod fly. The damage is caused only by the tiny maggots which bore into the pods and feed on the red-gram seeds. It is distributed throughout the red gram grown area of the India.

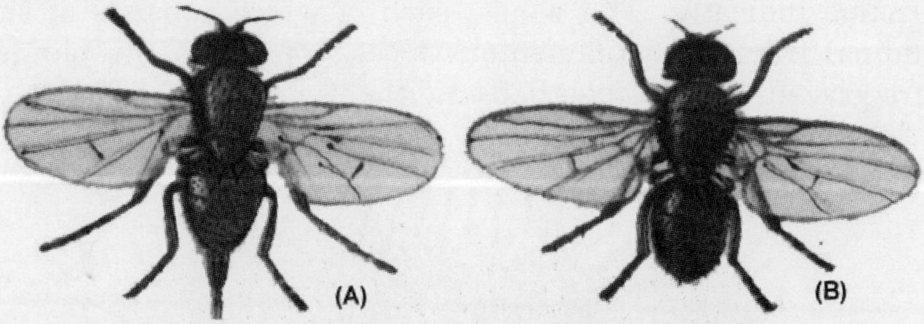

Fig. 65. Red gram pod fly *Agromyza obtusa* - **A.** Female, **B.** Male (From Pradhan, 1969)

Life history

The fly thrusts its minute eggs into the shell of a tender pod. The maggots hatch in about 3-4 days, and after feeding under the epidermis for some time they enters the seed. They become full grown in about 5-10 days. Pupation take place within the damaged pods. The pupal period lasts 7-11 or more days depending on the climatic conditions. The adult flies emerge by cutting holes. The entire life cycle is completed in about 9-30 days depending on the prevailing temperature. There are several generations in a year.

Control

The pest can be controlled by spraying the crop with 1.5 litres of endosulfan 35 EC or 750 ml monocrotophos 40EC in 1000 litres of water/ha or dusting with 5% carbaryl or 5% endosulfan.

---

* Same as *Melanagromyza obtusa* Mall.

5. TUR PLUME MOTH

Exelastis atomosa Walsh. (Lepidoptera : Pterophoridae) (Fig. 66)

This is a very specific pest which prefers only the red gram. It occurs in most of the regions of India growing red gram (Tur or Arhar)[1]. The damage is caused by the caterpillars. The caterpillar cuts a hole in the red gram pod opposite to a seed developing inside and then inserts its head in that hole and feeds from out side on the developing seed. When one seed is consumed, the larva leaves that hole and cuts another opposite, to another developing seed. Thus, it damages quite a high percentage of developing seeds which can not be replaced and are thus totally lost. The adult is a small (about 7 mm. in length and 15 mm. wing span), beautiful greenish brown plume moth, as its name indicates. The wings, each of which consists of several longitudinal fringe like filamentous lobes, constitute the plume and they are gracefully held together on either side at right angles to the body of the moth when it is at rest.

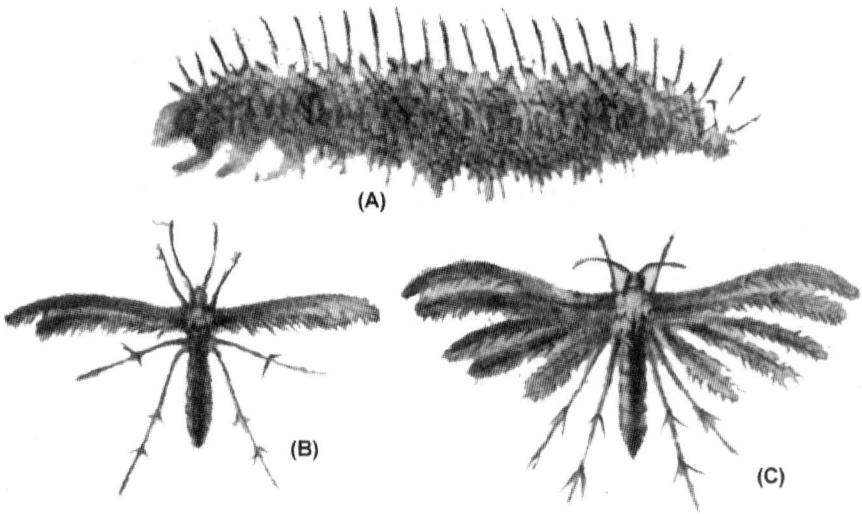

Fig. 66. Tur Plume moth *Exelastis atomosa* - **A.** Larva, **B.** Moth, in normal resting position, **C.** Moth (set) (From Lefroy, 1909)

Life history

The female moth lays about 17-19 eggs singly on the tender pods of the food plant. The eggs hatch into tiny larvae about 1mm in length

[1] Pigeon pea / *Cajanus cajan* Mills.

in about 2-5 days. The larvae feed on the pods and became full grown in about 2-4 weeks and measure about 7 mm. long. The colour of the larva is a variable combination of green and brown tinge. Pupation takes place outside the pod on its surface or in the entrance hole of the pod. The pupal period lasts from 3 days to more than a week and the whole life cycle is completed in about 15-40 days depending on the prevailing temperature.

Control

This pest can be controlled by spraying the crop with endosulfan 35 EC (2 ml in 1 litre of water) @ 800-1000 litres/hectare or dusting with 5% carbaryl. The treatment should be given when the pod formation starts and the larvae are found feeding them.

6. LENTIL POD BORER

Etiella zinckenella Treit. (Lepidoptera : Phycitidae) (Fig. 67A,B,C)

The Lentil pod borer is a serious pest of lentils as well as green peas both and is also found on a variety of other pulse crops in different parts of the country. The moths are grey with a wing expanse of 25 mm. The fore wings have dark marginal lines and are interspersed with ochreous scales.

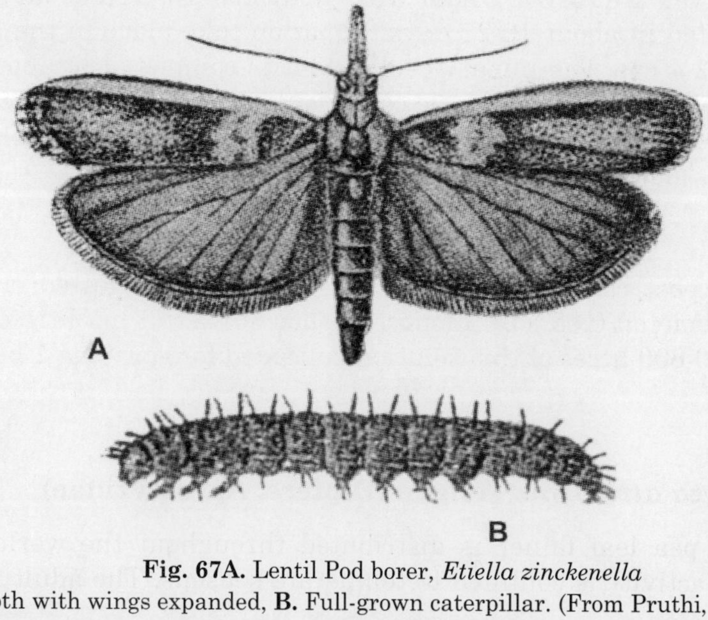

Fig. 67A. Lentil Pod borer, *Etiella zinckenella*
A. Moth with wings expanded, **B.** Full-grown caterpillar. (From Pruthi, 1969).

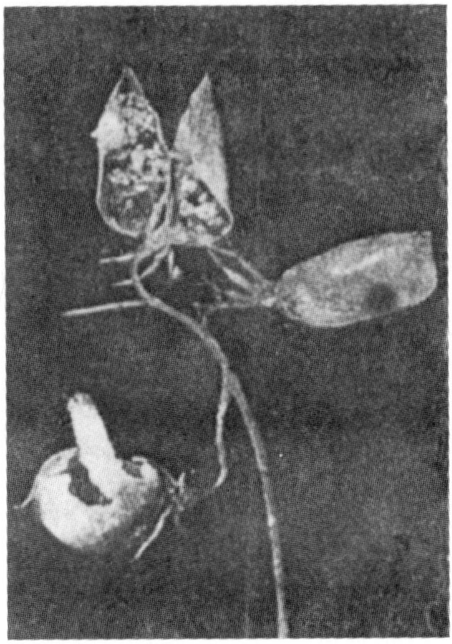

Fig. 67C. Damage caused by the Lentil pod borer. The borer is also seen in a pod (From Atwal, 1976)

Life history

The female moth lays 50-100 eggs either singly or in clusters on the various parts of the plants. The eggs hatch in about 5-35 days depending on the prevailing temperature. The newly emerged tiny greenish larvae enter the pods and eat away the young seeds. The full grown larvae are rosy in colour with a purplish tinge. The larval stage is completed in about 10-27 days. Pupation take place in the soil at a depth of 2-4 cm. The pupal development is completed in about 10-15 days depending on the climatic conditions, there after the moth comes out from the pupal case and starts the new generation. There are about 5 or more generations in a year.

Control

The pest can be controlled by spraying the infested crop with methyl demeton 0.05% or monocrotophos 40EC @ 1 ml/ litre of water. About 600-800 litres of this solution is needed for spraying 1 hectare.

7. PEA LEAF MINER

Phytomyza atricornis Meigen (Diptera: Agromyzidae)

The pea leaf miner is distributed throughout the world but in India its activity is confined to temperate seasons. The adult is a tiny fly, about 1.5 mm. in length and with a wing span of less than 4 mm. The flies are blackish-grey or silvery in appearance. The larvae are destructive and they make prominent zigzag silvery tunnels in the leaves. An intersting point about the habit of this fly is that it combines the action of feeding and oviposition. It punctures the tender leaves with its ovipositor and then turns round to feed on the plant juice exuding from the puncture. The males are devoid of ovipositor and therefore cannot puncture, wait in attendance near the female to lick the juice after the latter have left the place.

Life history

Each female fly makes numerous punctures on each leaf. The fly oviposits in the puncture, and lays one egg in each cavity. Each female fly is capable of laying about 350 eggs in its whole life span which may extend to near about one month. The eggs are extremely small, about 0.3 mm. in length and oval-elangate in shape. The eggs hatch within 2-5 days into tiny maggots which begin to feed on the mesophyl of leaf without damaging the two epidermal layers. This leads to the formation of zigzag silvery galleries. Within 4-12 days the larva develops to its full size of about 3 mm. and then pupates within its last larval skin which hardens in due course. The pupa becomes reddish-brown or dark-brown in colour. The pupal period ranges from 6-16 days after which the fly emerges to start another generation. There are several generations in a year.

Control

The collection of damaged portions of the plant when the attack is just started so that the attack can be nipped in the bud and further multiplication is checked. Poison baits against the adult flies have also been recommended. The pest can be controlled with spraying any one of these inceticides on the crop: dimethoate 0.02%, carbaryl 0.1% or dimecron 100 EC @ 250 ml/ha or metasystox 25 EC @ 1 litre/ha in 1000 litres of water. All the pests of pulses can be controlled by spraying endosulfan 35 EC @ 1 litre/ha or phosphamidon @ 250 ml/ha mixed in 1000 litres of water.

19

PESTS OF SORGHUM

1. SORGHUM SHOOT FLY

***Atherigona soccata Rond. (Diptera: Anthomyiidae)**

The sorghum shoot fly is a serious pest of sorghum, and it is widely distributed throughout the world. In addition to sorghum, it is also a pest of maize, wheat, rice and plants of grass family. It damages the seedlings and earlier crops. The maggots bore into the stem and feed on the basal parts of the young shoots causing a typical dipterous dead heart. The maggots cause a considerable loss in the yield of grains. The adult looks like a small grey housefly, about 3mm long. These flies have long adult life of about 1 month or more. They feed on honey dew and other sweet material available in nature.

Life history

About 40 eggs are laid singly by a female-fly on the lower surface of the leaves. The eggs measure about 1.5 mm. in length, are elongated and flattened, somewhat boat shaped and are provided with two wing like lateral projections. The eggs hatch in about 2-3 days. The maggots are extremely small which move down the leaves and pass between the leaf sheath and stem and bore into the stems of the seedling at about the level of the growing point. They feed on the developing shoots for about 7-10 days. Pupation takes place in the soil which lasts for about a week after which the adult fly emerges and starts laying eggs on other seedlings. The total life cycle takes about 15-32 days. There are several generations in a year.

Control

It can be controlled by planting the resistant varieties and by spraying 0.2% carbaryl or 0.05% endosulfan. Application of 10%

---

* Same as *Atherigona indicus*

phorate (thimet) granules in soil before sowing @ 15 kg/ha should be done. Two sprays of 0.01% methyl demeton (1 ml chemical in a litre of water) solution @ 560 litres/ha at 15 days interval after sowing controls the pest.

2. SORGHUM EARHEAD BUG

Calocoris angustatus **Leth. (Heteroptera : Miridae) (Fig. 68)**

It is one of the most serious pests of sorghum in India. The nymphs and adults suck out the juice from just developing sorghum grains which are in their milky stage. A continuous feeding by the nymphs and adults of these bugs results in the shrivelling of grains. When heavy infestation occurs the ears become black in the initial stage and finally they dry up hence the grains are not produced at all. In the adult stage, this bug is about 5-8 mm. in length, a little more than 1 mm. in breadth and yellowish green in colour.

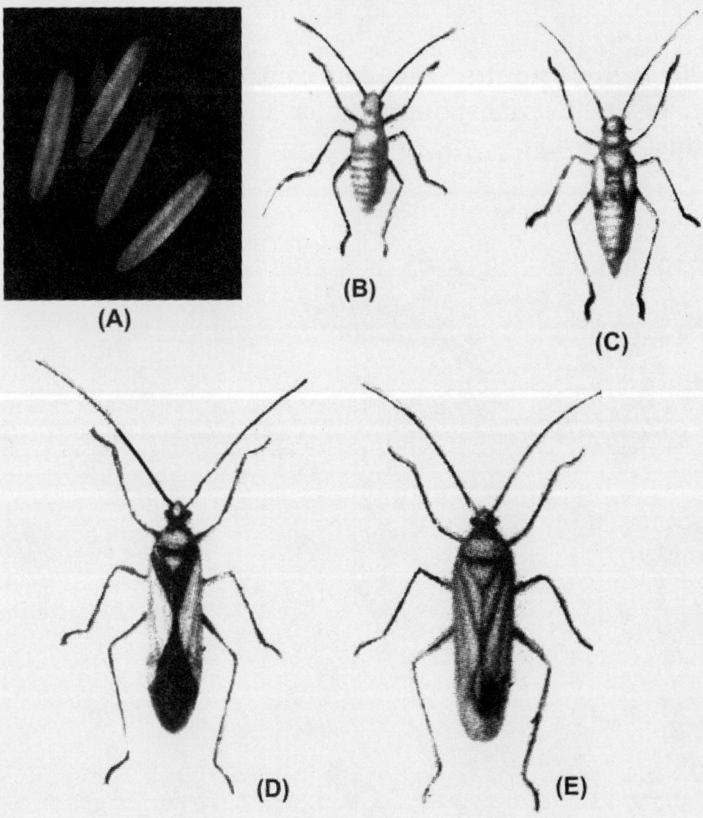

Fig. 68. Sorghum earhead bug *Calocoris angustatus*. **A.** Eggs, **B.,C** Nymph, **D.** Male adult, **E.** Female adult (From Pradhan, 1969)

Life history

The adults are attracted to the sorghum crop as soon as the earheads begin to make their appearance from the enfolding leaf sheath. They immediately begin to feed and lay eggs in the crop. A female bug lays as many as 150-200 eggs in between the anthers of the sorghum florets by inserting its ovipositor. The eggs are extremely small, cigar shaped about 1.5 mm. long. These eggs hatch out mostly in less than a week i.e. in about 6 days. As each insect lays between 150-200 eggs, the population of the pest increases by 75 to 100 times within about a week since the crop begins to attract the adults from the surrounding uncultivated grasses. The nymphs just after emergence start feeding on developing grains in milky stage. The nymphs pass through 5-instars and develops into adult in about 3 weeks. The life cycle is completed in one month under favourable conditions. There are several generations in the season.

Control

Nymphs and adults can be controlled by spraying 0.05% malathion or 0.03% phosphamidon or 0.1% carbaryl (W.P.) or 4% carbaryl dust.

20

PESTS OF CEREAL CROPS

1. JOWAR STEM BORER

****Chilo zonellus* Swinh. (Lepidoptera: Pyralididae) (Fig. 69 A,B)**

It is a major pest of maize and jowar. However with respect to maize it causes damage only to the young plants while all stages of jowar are severely attacked. The period of infestation comes from the middle of March and continue upto September with a peak during August. Normally the attack on the plants starts from the top with the caterpillars boring into the shoots at the upper end. However, there are instances where the caterpillar enters the shoot from the side, punctures even stems as high as 8 cm. from the ground. In either case they ingest the sap from the plants and bring about its withering which ultimately results in the formation of dead hearts. The withered plants have reduced leaves and stem. An attack by this pest from the top end of the plant is indicated by the presence of numerous punctures on the leaves forming the whorl.

The adult insect is a yellowish brown moth 20-30 mm. across the wing with deep brown spots, long legs and two projecting palps in front of the head which give the appearance of a beak. Sexual dimorphism is characteristically exhibited in this species. The hind wings of male have a smoke like or pale straw colour, while those of females are bright whitish. Further, the abdominal tip of males is smooth and bluntly pointed while those of females is broad and covered with a thick tuft of hairs. The adults are short lived.

---

* Same as *Chilo partellus* (Swinh.).

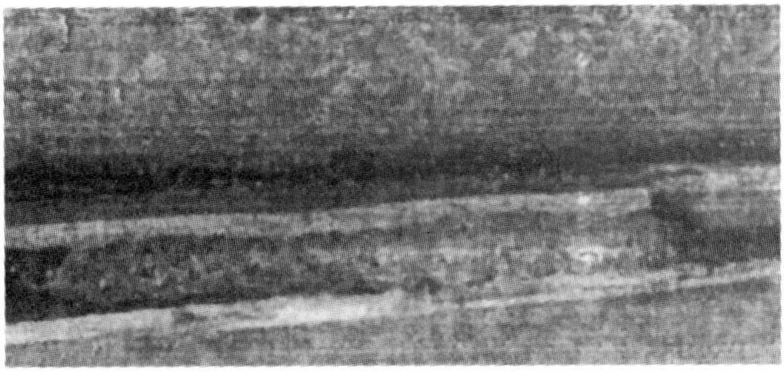

Fig. 69 A. Jowar stem borer *Chilo zonellus* (inside the stem)

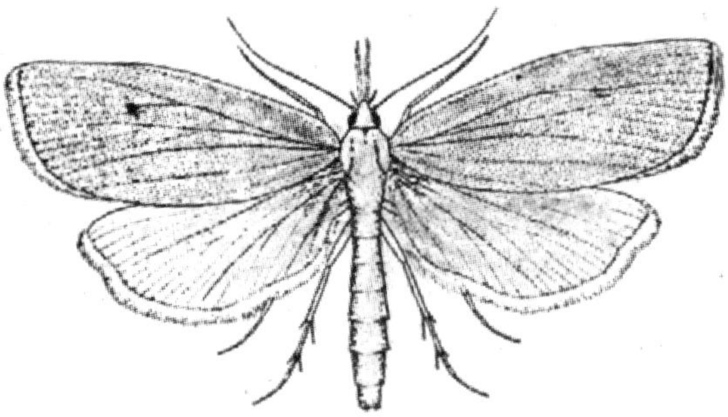

Fig. 69B. Adult female of jowar stem borer *Chilo zonellus*.

Life history

The female lays greenish or creamy white, flattened, oval about 0.8 mm. long eggs at night on the lower surface of the leaf near mid rib in two rows overlapping each other. The eggs are normally laid from April to May. Each row contains about 50-100 eggs which in course of time change their colour first to orange and then to yellow. Hatching occurs in about 7-10 days liberating young caterpillars. These immediately feed on leaves, on which they have hatched and then gradually penetrate into the leaf at the mid rib region to start their journey towards the main stem for the principal attack. Often their course from the leaf to the stem is interspersed with small punctures on the leaf for the purpose of ventilation. Thus they reach the shoots tunnelling and boring their way from the leaf. Soon the main stem looses its strength and becomes a dead heart. The caterpillar thus draws its nourishment from the shoots and become fully matured in about 3-4 weeks. At this stage it prepares itself for pupation. It immediately makes an exit hole in the shoot and undergoes pupation within the infested stem in a chamber specially lined with thread like secretion. The pupal period lasts from 7-10 days. The pupae are shining brown in colour and possess six spines arranged in two rows at the tip of the abdomen. At the end of pupal period fresh moth emerges through the exit hole. The entire life cycle lasts for about 5-6 weeks. Generally there are two broods in a year.

Control

Many precautionary measures such as thorough examination of field before planting, removal of stubble trash and deadhearts from the plough field are employed. Besides these certain electrical devices such as attraction of moths by light traps are also adopted. Often biological control measures have been found to be successful. These include the introduction of egg parasites such as *Trichogramma minutum* and *Bracon chinensis.* Spraying of 1.5 litres thiodan 35 EC or 2 kg (50% soluble) carbaryl in 1000 litres of water/ha. 15-20 days after sowing helps in controlling the pest.

2. GUNDHI BUG

******Leptocorisa varicornis*** Fabr. (Heteroptera: Coreidae) (Fig. 70)**

It is a common pest of rice and is a major one. Both the nymphs and adults destroy the ripening grains of the plants usually during August to October, extracting their milky juice with the consequence that the ears turn white forming shrivelled grains and preventing their maturation. A serious attack results in the appearance of empty, dry, brown earheads on the plant. Yield losses of 10-40% are common and in severe infestations the entire crop may be destroyed. Plants such as those of jowar, maize, bajra, etc. form the secondary host of this insect.

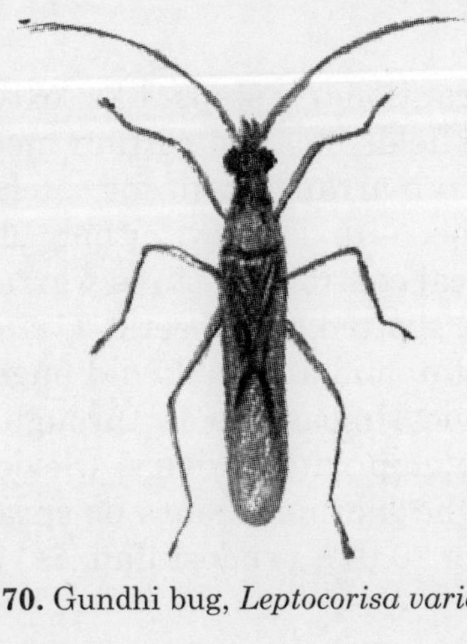

Fig. 70. Gundhi bug, *Leptocorisa varicornis*

* Same as *Leptocorisa acuta* (Thunb.)

The adult insect is slender, green or light brown in colour. It emits strong upleasant aromatic odour and hence the name 'Gundhi bug'. Both male and female individuals are identical in appearance except that the abdominal apex of the latter is more swollen than that of the former.

Life history

The female deposits eggs in one to three rows on the leaves of the paddy plant, each row containing 5-20 eggs. The eggs are black, oval and flattened but concave from above. Each egg has a dark brown shell with a white spot indicating the point of rupture during hatching. The eggs are fixed to the leaves by a gummy substance. The eggs hatch into young, pale green nymphs in about 6-8 days. These nymphs are devoid of wings but green in colour in the beginning but possess long legs and antennae. Slowly the wing also develop and the nymph moults 5-times with in a period of 17-27 days before attaining the adult condition. An important feature of the final moult is the transfer of functional odouriferous glands from the abdomen to the thorax and the formation of three tarsal joints. When the plant is at a tender stage both nymphs and adults congregate in large number around ripening ears, tender shoots and leaves. The entire life cycle is completed in about 4-5 weeks and in a year there are generally 5 broods during the paddy season. Multiplication normally occurs from July to November, hibernation from December to February and their population on grasses from March to June.

Control

Besides many precautionary steps like extermination of wild grasses from the sowing fields etc. and certain mechanical (electrical) measures such as light trap arrangement for catching the insects, the general tendency has been to employ either biological control or chemical control. Biological control measures warrant the introduction of certain beetles like six spotted tiger beetle, *Cicindela – sixpunctata** or certain species of Pentatomid and Reduviid bugs which prey on the adults. Control by chemical measures is through the application of insecticides. Dusting the crop with carbaryl (Sevin) 5% or malathion 5% dust @ 20 kg/ha in the morning hours or spraying the crop with 0.05% monocrotophos or 0.07% endosulfan is useful. Dusting is

*Same as *Cicindela sexpunctata* Fabr.

usually done on the earheads, especially when they are in the milky stage and the bugs appear in large numbers.

3. RICE GRASSHOPPER

***Hieroglyphus banian* Fabr. (Orthoptera: Acrididae) (Fig. 71)**

It is a serious though sporadic pest of eastern U.P., damage to the paddy is caused both by the young and adult stages of the insects which feed upon the leaves and soft ear grain of the plants. The insects also attack other plants like sugarcane, jowar and different types of grasses.

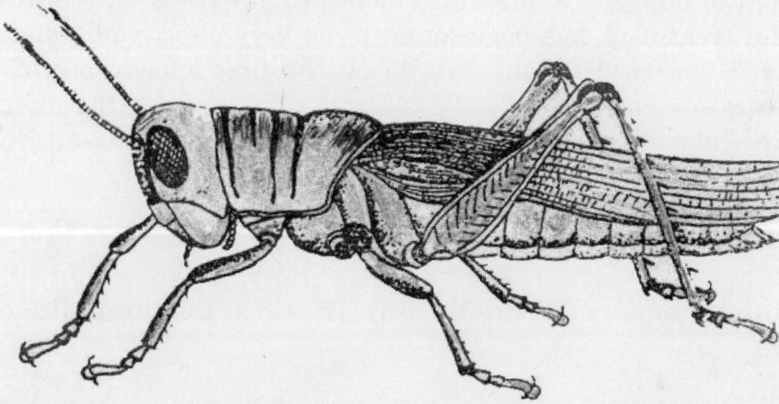

Fig. 71. Rice grasshopper, *Hieroglyphus banian.*

The adult is a green, dry grass coloured medium sized individual with long slender legs. The young nymphs are very much like the adults except that they are smaller in size and possess a brownish yellow colour and allied species of this genus *H. nigrorepletus* Bol. is a minor pest of jowar.

Life history

The female lays yellow eggs resembling rice grains in clusters usually during October month. The eggs are closely packed in hard "cement" which forms a pod like cylindrical structure termed "egg pod". Each pod containing 30-40 eggs is laid 5-8 cm. deep in the soil by inserting the abdomen. The eggs remain burried under soil during hibernation stage i.e. upto June.

The eggs hatch during the monsoon season into young nymphs which immediately starts feeding on nearby grasses and later on migrates to the main crops. These nymphs become full grown in about 8-10 weeks, during which period they moult 5-times (sometime 6 times) at an interval of 10-15 days each. The young nymphs known as hoppers resemble very much the adult except for the fact that these hoppers are devoid of wings. The nymphal period on the whole extends from 10-12 weeks after which a fully developed adult is formed. The entire life cycle lasts for about 17 weeks. Generally there is one brood in a year which appear during the monsoon rain.

Control

As in the case of other pests, certain measures of success is achieved by adopting physical methods such as hand collecting and destroying the hoppers, and arrangement of light traps etc. However, insecticidal treatment has been found to be very effective by dusting 12.5 kg. 10% carbaryl (sevin) dust/ha on the first appearance of the pest. The grasshoppers can also be controlled by spraying the infested crops with endosulfan 1250 ml or 875 ml methyl parathion 50 EC in about 1000 litres of water per hectare.

4. PADDY GALL FLY

Pachydiplosis oryzae (Wood-Mason) (Diptera: Cecidomyiidae) (Fig. 72)

The paddy gallfly is very small in size and has a wing span of less than 2mm. The fly has long slender legs and produces in the paddy plants long tubular galls of silvery appearance in the place of normal stems with earheads. Hence the crop attacked by this insect is said to be suffering from "silver-shoot" desease. Its distribution covers East and South Asia. In India, it attacks the paddy crops with varying intensity in most of the paddy growing areas but its infestation is very serious in the costal regions of the country on account of weather conditions. The larva of the fly is known as maggot and is very harmful. Feeding of the maggot produces galls which kills the growing shoots.

Life history

The fly breeds on a number of wild grasses and it is attracted to the paddy crop at its tillering stage. The female fly which is pink in colour lays about 200-300 reddish elangated eggs about 0.5mm. in length generally on the lower surface of the basal portion of the leaf. The tiny larva hatching out within 3-5 days creeps down between the leaf-sheaths till it reaches the apical point of the central shoot or tiller, as the case may be, and there it enters the tissue and destroys the growing point. As a result of which the normal growth stops and a small gall is produced. This gall gradually elongates and a long hollow shoot of silvery appearance emerges in the place, which in a normal plant is occupied by the stem bearing the earhead. Thus the earhead formation is totally prevented.

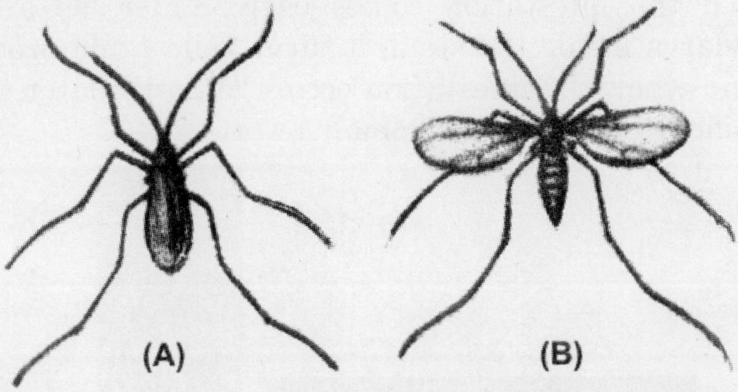

Fig. 72. Paddy gallfly *Pachydiplosis oryzae*, **A.** Adult fly in sitting and **B.** flying attitudes.

In the meantime, the larva becomes full grown in about 10 days and pupates inside the tubular gall. The pupal period lasts about 3-5 days and thereafter, the fly emerges out of the gall through a hole already made before the pupal stage. Thus, the whole life cycle is completed in about 3 weeks. The adult fly lives only for about 3-4 days and therefore just after emergence it starts a new generation. In favourable climatic conditions there are 6-8 generations in a year.

Control

The paddy gall fly infestation can be prevented, by spraying the crop with a number of insecticides, some of them are : phosphamidon 0.05%, diazinon 0.05%, carbaryl 0.2%, dimethoate 0.02%, lindane 0.5%. The collection and destruction of "silver shoots" can also help, provided it is started early in the season and the galls are rouged out as soon as these are seen.

5. PADDY STEM BORER

***Tryporyza incertulas* (Walk.) (Lepidoptera : Pyralididae) (Fig. 73A)**

It is interesting to note that unlike other lepidopterous larvae, this paddy stem borer is "highly monophagous" and no alternate host plant besides rice has been recorded so far. It is found in practically all the rice growing regions of India as well as many rice growing tracts of the world. The adult moths have the wing expanse of 25-40 mm. and the wings are ochrous yellow in colour. In the female moth, there is a prominent black spot on each of the forewings while this spot is indistinct in males. The female moth is bigger than the male. The female moths have a tuft of brown-yellow hair at the tip of their abdomen. Regarding the harm done to the paddy plant, the whole seedling dies if the infestation comes early. Later, as a result of feeding of the larva inside the stem, it turns yellow and produces the dead heart, and when the infestation occurs at a still later stage, the earheads become white and grain remains worthless.

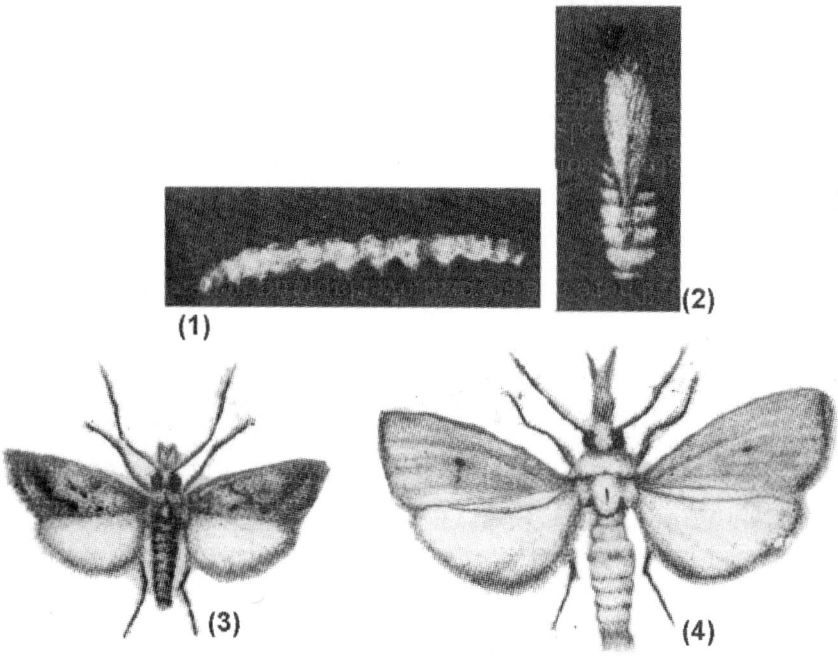

Fig. 73A. Paddy stem-borer *Tryporyza incertulas* **1.** Caterpillar, **2.** Pupa, **3.** Female moth, **4.** Male moth (From Pradhan, 1969)

Life history

The eggs are laid in 2-6 batches and each batch has about 65-85 eggs, on the under side of the leaves and are covered by buff coloured hairs contributed by the anal tuft of the female moth. The first stage larvae hatching out from the egg masses in about 6-9 days, crawl about on the plant or hang down from the leaves by silken thread and bore into the stem and enter the tissue for the rest of their life as larvae and pupae. The larva passes through 4-6 instars and is full fed in about 15-30 days and attains the length of 2-2.5 cm. The full grown larva has a smooth surface of yellowish green tinge. The head is orange yellow. Before pupation it cuts open another hole which is located above the water level in the field through which ultimately the moth comes out. After the pupal period is over which lasts about 8-12 days depending on weather, the moth emerges and marks the beginning of the next generation. The whole life cycle is completed in about 30-50 days. There are several overlapping generations in a year.

Control

At the egg stage, collection and destruction of egg masses is very effective. The pest can also be controlled by spraying the crop with any one of these insecticides: methyl parathion 0.08%, phospha-midon 0.05%, diazinon 0.06%.

6. PADDY LEAF MINER

***_Hispa armigera_ Oliv. (Coleoptera: Hispidae)** (Fig. 73B)**

The grubs are leaf miners. Also knwon as Rice Hispa. This is also a pest of paddy which has been found widely distributed in this country and abroad. The damage is caused by the grubs and adults both which feed on the green delicate parts of the paddy leaves. The adult beetles are blue-black in appearance, measuring about 5 mm in length and are recognized by numerous short spines on the body.

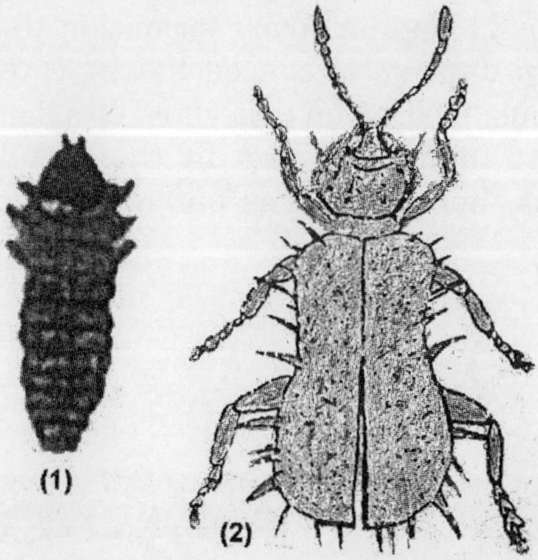

Fig. 73B. Paddy leaf minor, *Hispa armigera* **1.** Grub, **2.** Adult

* Same as *Dicladispa armigera* Oliv.

** Originally treated as subfamily of chrysomelidae.

Life history

Each female individual lays about 50 to 60 eggs inside the leaves. The tender parts of the leaves are preferred for laying the eggs. After about 5-days, yellowish grub emerges which feeds inside the leaf-mine for about 10 days. Pupation takes place inside the mine and adult emerges in about 4 to 6 days. There are several generations in a year.

Control

For controlling this insect spraying of 0.5% of insecticides like methyl parathion, and phosphamidon was found useful. Spraying the crop with 0.05% monocrotophos, as well as 0.07% endosulfan at the time of severe infestation also controls the pest-population.

7. SWARMING CATERPILLAR

***Spodoptera mauritia* Boisd. (Lepidoptera : Noctuidae) (Fig. 74A)**

Swarming caterpillar is a sporadic pest but is very serious indeed whenever it appears in large number. It is distributed in the entire costal region of India as well as in the Oriental and Australian regions. As the name indicates, in caterpillar stage this pest has a tendency to migrate from field to field in large swarm. The swarm browses a field like cattle and when field is complitely finished, the swarm marches on in regular army formation to the adjoining field. Thus a very heavy damage is caused by the caterpillars. The caterpillar is polyphagous, feeding on cultivated crops and wild grasses, but paddy is the most important crop in the costal areas due to the weather conditions, hence, it terns out to be a very serious pest of paddy.

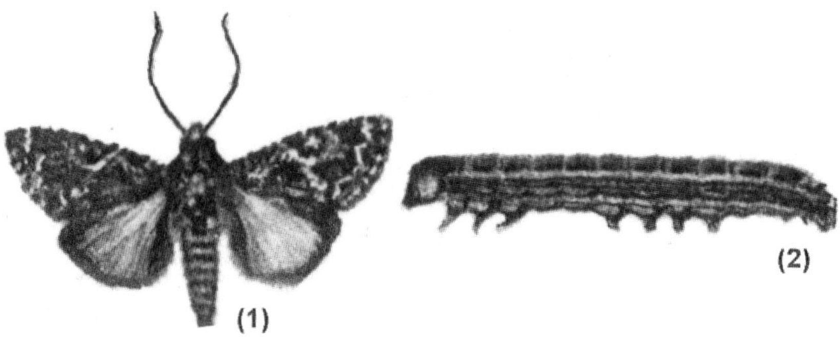

Fig. 74A. Swarming caterpillar, *Spodoptera mauritia* **1.** Adult, **2.** Caterpillar.

Life history

The adult is a medium sized moth of grey brown colour with beautifully designed patchy forewings which have a wavy border pattern on the fringe. The hind wings are practically white. The moth measures about 15-20 mm. in length and about 38-40 mm. across the wings. These moths lay eggs on the under side of the leaves in large masses, each containing up to several hundred eggs and about 1000 eggs are laid by a single female in her life time. They are covered with buff coloured hairs from the female's body. The eggs hatch in about 7 days or so depending on the prevailing temperature. The young larvae immidiately begin to feed on nearest available plant. The larval stage lasts about 20-28 days. The first stage larva is extremely small which gradually grows through several moults into a full grown larva of about 3-4 cm. in length. The colour pattern of the larvae consists of longitudinal stripes which in the beginning have a mixture of green and yellow with a reddish stripe along the side and which later on are mostly of yellowish brown hue. Pupation take place in the soil. The pupal stage lasts from 10-12 days after which the moths of the next generation appear and give rise to a new brood. Both the moths and caterpillars are nocturnal. There are varying number of generations in the different regions of the country.

Control

For preventing the damage caused by the swarming caterpillar, the collection and destruction of the large egg masses, is very useful. Regarding chemical control, stomach poisons or a contact + stomach poison is very effective. Dusting the crop with sevin 10% or spraying 0.04% endosulfan or 0.2% carbaryl is also very effective.

8. PINK BORER

Sesamia inferens Walk. (Lepidoptera: Noctuidae) (Fig. 74B)

The pink borer is a polyphagous insect and is distributed throughout India. It is a serious pest of maize and rice. Its larvae are pinkish-brown in colour which bore down the stem making long tunnels throughout the whole length. The central shoot is completely destroyed and dead hearts are formed. It feeds on wheat, barley, paddy, sugarcane, jowar etc.

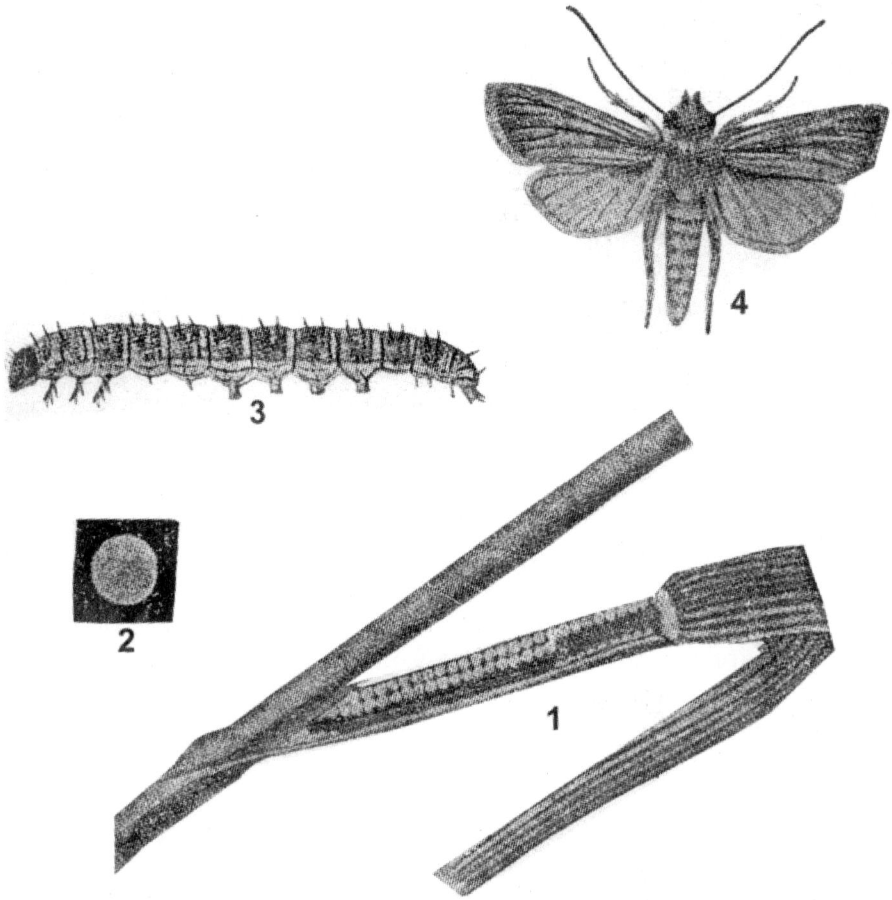

Fig. 74B. Pink borer, *Sesamia inferens.* **1.** Eggs laid in the leaf-sheath, **2.** An egg (highly magnified), **3.** Full-grown borer, **4.** Moth in flying posture. (From Pruthi, 1969)

Life history

The pest breeds actively from March-April to November on rice and then migrates to the wheat crop. The moths are nocturnal and lay 150 eggs on leaves or on the ground. The eggs hatch in about 6-8 days and the young caterpillars bore in to the epidermal layers of the leaf sheaths. Later on, they bore into the stem as a result of which the growing shoots dries up producing dead hearts. When the attacked plants dry the larvae move on to adjoining plants, they are full fed in 3-4 weeks and pupate inside the stem or in between the stem and leaves. The pupal stage lasts about a week and the life cycle is completed in about 6-7 weeks. There are 4-5 generations of the pest in a year.

Control

Spraying methyl parathion 0.08%, phosphamidon 0.05% or diazinon 0.06% is effective in checking the infestation of this insect.

9. ARMYWORM

Mythimna separata Walk. (Lepidoptera: Noctuidae) (Figs. 75, 76)

It is a pest of wheat, maize, jowar, bajra, sugarcane etc. The caterpillars are voracious, eating entire leaves and the whole plant, usually during nights and migrate from one filed to another field.

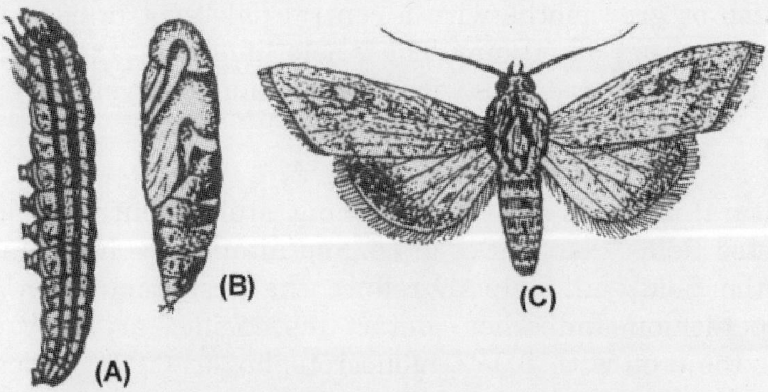

Fig. 75. Armyworm *Mythimna separata* **A.** Caterpillar, **B.** Pupa, **C.** Adult (From Atwal, 1976).

Fig. 76. Wheat plants exposes heavy infestation of armyworms attacking the crop. Note how leaves are being consumed (From Pfadt, 1962).

Life history

Eggs are laid either singly or in clusters of about 100, inside the rolled leaves or between the leaf sheath and stems. They are subspherical, greenish white, turning yellow and take about 4-13 days or more to hatch, which depends on the climatic conditions. Older caterpillars are gregarious and have voracious appetites. The caterpillars vary considerably in colour, from green to pink but have four distinct longitudinal black stripes and reach a mature size of 35-40 mm. after about 14-22 days in spring or 80-100 days in winter. The pupa is dark brown in colour 15-19 mm. in length. Pupation takes place 4-5 cm. deep in the soil. Pupal stage is completed in about 7-22 or even more days. The length of pupal stage depends on temperature. The adults are reddish or grey moths with a central pale spot in each forewing. Wing span is from 35-50 mm. The whole life cycle takes about 25-64 days (31 days on an average) with 5 generations per year.

Control

Cultural methods such as ploughing, stubble burning, flooding of the infested fields, removal of grass and alternative host plants from around the fields, all help to reduce the pest population. Dusting methyl parathion and other contact insecticides are very effective. Spraying the crop with 0.05% monocrotophos or 0.07% endosulfan at the time of heavy infestation reduces the pest population.

10. WHITE GRUB

***Holotrichia consanguinea* (Blanch.) (Coleoptera: Melolonthidae) (Fig. 77)**

The grub feeds on the fibrous roots of maize, sorghum, brinjal, sugarcane etc. and is found in the upper layer of the soil about 5-10 cm. The full grown grub measures about 35 mm. long, is white in colour with brown head. The beetles are dull brown and about 18 mm. in length. While the grubs feed on roots, the adult beetle feeds on the plant leaves. It has also been reported as a major or minor pest of certain fibre crops, like many other insects, hence, considered as a pest of several crops and is polyphagous.

Life history

The adult beetles lay eggs singly upto a depth of 10 cm. which hatch in about 7-10 days. The development completes in about 10 weeks. After the monsoon the grubs pupate in the soil. The pupal stage lasts for about 15 days. The beetles remain in soil at a depth of 10-20 cm. and come out only at night for feeding. The adults remain in soil and undergo diapause for about 8 months. They remain active only in rainy season i.e. June/October. There is only one generation in a year.

Fig. 77. A white grub feeding on corn root

Control

The white grubs can be checked by applying 10% carbaryl (Sevin) dust @ 12.5 kg/ha in the field. The soil can be treated with phorate granules, and quinalphos to control this pest. It may be pointed out that a number of pests of cereals like stemborers, aphids, hairy caterpillars, grass hoppers, and earhead bugs etc. can be controlled by using diazinon 5% or endosulfan 4% granules @ 20-25 kg/ha.

21

PESTS OF FRUITS AND FRUIT TREES

1. ANAR BUTTERFLY

Virachola isocrates Fabr. (Lepidoptera: Lycaenidae) (Fig. 78)

The anar butterfly is distributed throughout India and is a serious pest of pomegranate. Besides anar, the list of food plants includes guava, ber, tamarind, apple, orange etc. Damage is caused by the larva which bores into the fruit and feeds on its contents. From the wound caused by the larva, offensive smelling matter oozes out as a result of which the fruits are infected with a number of bacteria and fungi.

Life history

This butterfly is very small and is metallic purplish brown in colour. The hind wings have tail lobes. They have also an orange patch on the fore wings and black spots on the hind wings. The female butterfly lays eggs singly on various parts of the shoot but the young caterpillar which hatches out within 7-10 days, bores into the fruit. The larval period lasts from 14-42 days or more depending on the climatic conditions. Pupation takes place either on the stem or in the fallen fruit. The pupal period varies from 7-30 days or more. The breeding of the pest continues throughout the year depending on the weather conditions.

Control

Collection and destruction of all infested fruits in the beginning of the fruiting season is very effective to control this pest because the larva directly bores into the fruits. Efforts should be made to lure away the adult butterflies. Spray with 0.05% endosulfan or 0.1% carbaryl is also very effective.

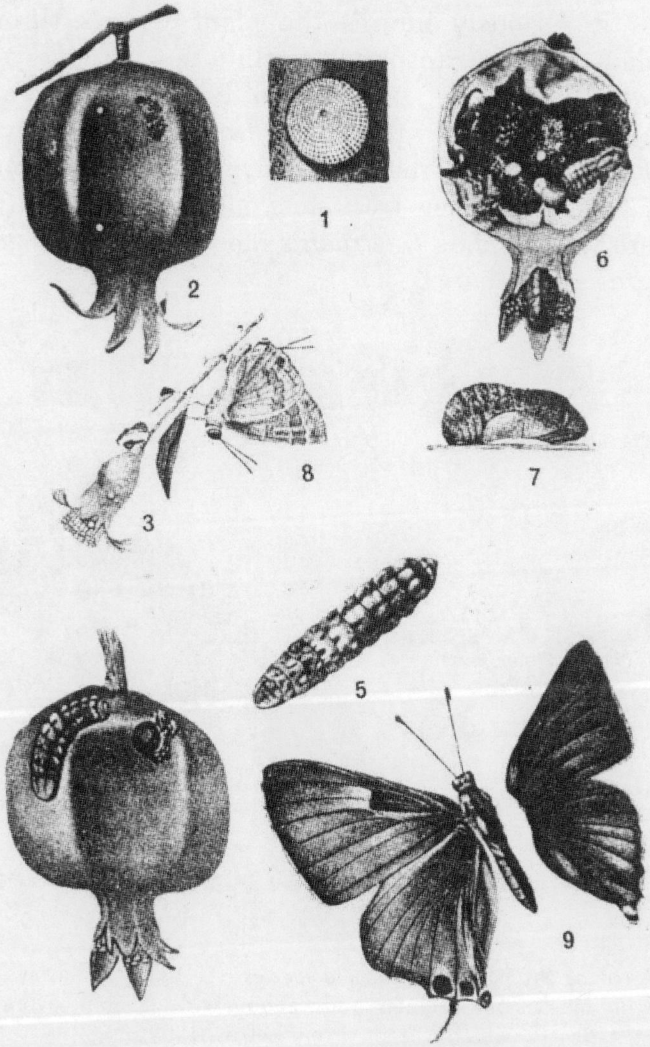

Fig. 78. Anar Butterfly - *Virachola isocrates,* **1.** Egg, **2.** Fruit showing two eggs on it and a hole through which the young caterpillar has gone in **3.** Flower with two eggs on it. **4.** Damaged fruit with a nearly full-grown caterpillar on it. **5.** Full-grown caterpillar, dorsal view, **6.** Damaged fruit cut open showing pupa inside **7.** Pupa, **8.** Butterfly in repose, **9.** Female butterfly with wings expanded, wings on the right are those of the male (From Pradhan, 1969)

2. BARK EATING CATERPILLAR

Indarbela species (Lepidoptera: Metarbelidae) (Fig. 79)

The list of trees attacked by the *Indarbela* is very large, some of them are mango, guava, pomegranate, jamun, citrus, ber etc. The damage is caused by the caterpillar which feeds on the bark and in this process it seriously injures the plant vessels through which nutritive plant sap is transported within the plant's system. As a result of which the tree growth and fruit bearing capacity are adversely affected. At times the infested branches can dry up and in cases of severe attack, the whole tree may die. It is a peculiarity of this pest, that it prefers older trees to younger ones. The species of *Indarbela* are identified as *I. tetraonis* Moore, *I. quadrinotata* Walker and *I. humeralis* Thunberg.

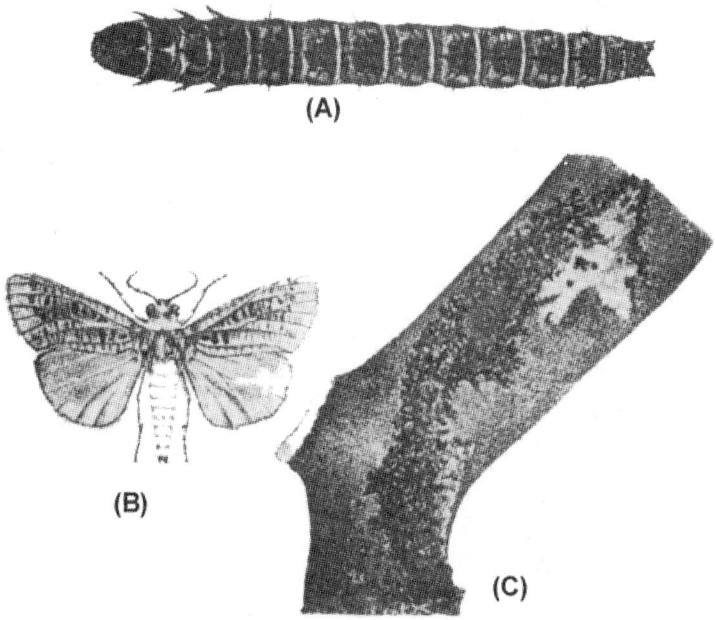

Fig. 79. Bark eating caterpillar *Indarbela species* - **A.** Larva full grown, **B.** Moth, **C.** Branch showing masses of excrement and webbing with which it covers the bark on which it feeds (From Lefroy, 1909).

Life history

The adult is a large sized moth with a wing span of about 4 cm. in female and about 3 cm. in male. They have light grey to light brick red colour provided with dark brown patches or dots. Each female lays a very large number of eggs but in small groups of 15-20 eggs each. The egg laying takes place in May and June and the incubation period is about 10 days. The caterpillar soon after emergence begins to feed on the bark and prepares the web under which it lives. The larval period is very long about one year. During this long period, it grows slowly. The full grown larva is about 4 cm. long. The larva is nocturnal. Pupation takes place within the larval gallery and the pupal period is only 3-4 weeks. At the end of the pupal period the pupa wriggles up to the opening of the gallery wherefrom the moth emerges. The life of moth is quite short and it completes egg laying within 2-3 days, after which it dies.

Control

The pest can be controlled by cleaning the web and fumigating the hole with ordinary fumigants like carbondisulphide or aluminium phosphide. Likewise inserting the cotton wool dipped in the liquid fumigant and thereafter plugging the hole with mud is very effective. The galleries made by the pest also can be treated with suitable insecticidal solution(s).

3. APPLE WOOLLY APHID

Eriosoma lanigerum **(Haus.) (Homoptera: Aphididae) (Fig. 80)**

It is a very harmful pest of apple and pear in India. Since a kind of white powdery secretion is produced by numerous glands on the body of the aphid which gives a general appearance of cotton-wool covering; hence, it is called "woolly aphid"; the adult body colour of which is actually purplish. The damage is caused by sucking the cell sap from all parts of the plant like the trunk, stem, branches, twigs, fruit stocks, leaf petioles etc. The infestation is associated with the formation of various types of deformities like large knots, galls etc. both on the underground and above ground portions of the plant. Other symptoms from desapping of the plant, like sickly appearance, stunted size, loss of vigour etc. are also produced and the fruiting capacity of the tree is very much reduced. As a result of severe infestation and disintegration of the roots, the trees get uprooted by wind.

Life history

The life history of this pest varies in details from place to place, depending on the climatic conditions. The pest overwinters either as eggs or immature nymphs on the under ground portions of the host plant. The eggs hatch and the nymphs mature during spring. Reproduction in this season is by parthenogenesis and vivipary. About 100 to 116 nymphs are produced by a female in her life time. The nymphs soon settle down to feed on the plant sap and within 24 hours thereafter begin to secrete the woolly filaments of wax over their body.

They undergo 4 moults before they become adult. The duration of nymphal period is about 10 days in favourable conditions and near about 100 days when it is very cold. During the whole of the summer and the mansoon, the reproduction is very rapid and both the winged and wingless forms are found. When winter approaches, the sexual forms appear, mate and lay eggs. Also, at this time there is migration from the aerial region to the root region for hibernation there in nymphal stages. There are numerous generations in a year.

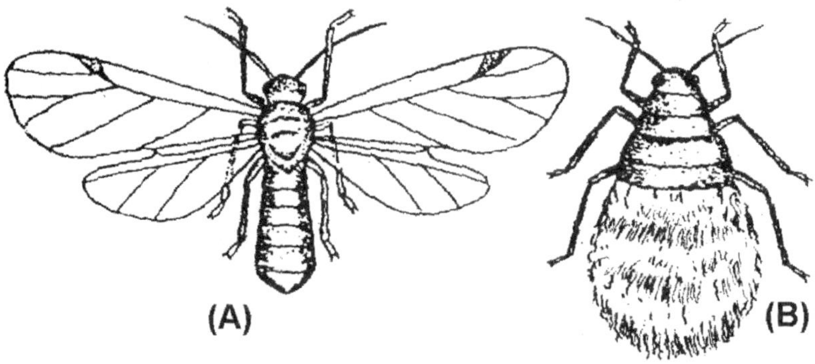

Fig. 80. Apple woolly aphid *Eriosoma lanigerum*, **A.** Winged male, **B.** Wingless woolly female (From Atwal, 1976)

Control

Like other aphids, *E. lanigerum* has a number of natural enemies in the form of parasites and predators. It can be controlled by introducing an exotic parasite, *Aphelinus mali* which attacks the last stage nymphs. The pest can also be controlled by spraying nicotine sulphate 0.04% or malathion 0.08% or 0.03% dimethoate. It has been suggested that fumigation against the root forms with paradichloro-benzene in a 15 cm. deep trench dug round the infested apple tree about 2 metres from it, is very useful. Further, insecticides should not be used in the areas where parasite *A. mali* is present.

4. LEMON BUTTER FLY

Papilio demoleus Linn. (Lepidoptera : Papilionidae) (Fig. 81)

The lemon butterfly is widely distributed in India and several other countries. It is not only a pest of citrus but of all the plants belonging to family Rutaceae. The young larvae feed on the tender leaves while the older ones on the mature leaves. As a result of heavy infestation the entire plant may be left without leaves and bears no fruits.

Fig. 81. Lemon butterfly *Papilio demoleus* - **1.** Egg laid on leaf (see: arrow), **2-6.** Larvae in various stadia, **7.** Larva suspended for pupation **8.** Pupa, **9 & 10.** Butterfly. (From Fletcher, 1914)

Life history

A female butterfly usually mates once and lays about 75-120 eggs within 2-5 days. A comparatively smaller number of eggs are laid in summer. The eggs are scattered singly on young leaves and tender shoots. The freshly laid eggs are pale green which become dark grey before hatching. The hatching takes place in about 3-4 days in favourable conditions and 5-8 days under unfavourable conditions. The larval life lasts 8-16 days in favourable conditions and 4 weeks under unfavourable conditions. A full grown larva is bright yellowish green measuring about 40 mm. long and 6.5 mm. wide. The young larvae are provided with white and black or brown markings, thus appear like the droppings of bird. Pupation takes place on the twigs or sticks. The pupal period lasts about 8 days in favourable conditions and 9-12 days under unfavourable conditions. The butterfly is fairly large and beautiful, measuring 94 mm. from wing tip to wing tip. The head and thorax of the butterfly are black and the underside of the abdomen is creamy yellow in colour. The wings are dull black which are ornamented with shiny yellow markings. One complete generation may take as little time as 2½ weeks and as much as 4½ months. There are 3-4 generations in a year and the winter is passed in the pupal stage.

Control

The larvae can be controlled by hand picking and by spraying the plants with 0.05% malathion, or 0.05% endosulfan or 0.1% carbaryl.

5. CITRUS RED SCALE

Aonidiella aurantii (Maskell) (Homoptera: Diaspididae)

The citrus red scale is one of the most destructive pest of citrus plants and fruits. It infest all the above ground parts of the plant. The leaves, branches and fruits may be covered with them. The pest feeds on the plant juice and devitalize the plant, and when the infestation is severe the leaves turn pale. These scales are about 2 mm. in diameter.

Life history

The pest is active throughout the year. The female scale insects give birth to nymphs. The first stage nymphs or the crawlers have well developed legs and antennae. They cover themselves with a white waxy secretion. The female nymphs moult twice at an interval of 10-

20 days, lose their legs and antennae and their cast skins are incorporated into the waxy covering which become circular, depressed scales. They reach sexual maturity in about 10-15 weeks, they are without wings and live for several months. The male scales, on the other hand are elongated and develop into winged adults in about 1-2 months. They fly around and fertilize the sedentary females. The pest passes through several generations in a year.

Control

The scale insects can be controlled by spraying the trees with phosphamidon 0.03% or dimethoate or malathion 0.05%. The sprays have to be repeated several times and special care should be taken while spraying methyl parathion.

6. CITRUS LEAF MINER

***Phyllocnistis citrella* Staint. (Lepidoptera : Gracillariidae) (Fig. 82)**

This insect is widely distributed in India and is a serious pest of citrus. It is also a pest of several other plants. The damage is caused by the larvae by making zigzag silvery mines in the young leaves. Due to the extensive mining by the pest, the leaf suffers badly, gets deformed and irregularly curled up in shape, unhealthy in look and defective in its function and finally it dries and falls off. The attack is comparatively much greater in leaves of fresh growth, young seedlings and varieties having soft succulent leaves.

Fig. 82. Citrus leaf miner *Phyllocnistis citrella* - **A.** Mined citrus leaf, **B.** Adult

Life history

The moths lay tiny, flattend, transparent eggs on the under side of the young leaves. The eggs are laid singly which hatch in about 2-7 days. Soon after emergence, the leg less larva enters the leaf tissue and begins to feed inside it and mine the leaf lamina. The larva is full fed in about 5-25 days. The full grown larva measures about 5 mm. in length and is yellowish green in colour. Pupation takes place within the galleries. The pupal stage lasts about 5-21 days after which the moth comes out. The adult stage is a tiny greyish moth with a wing span of 8-10 mm. The fore wings are white with two narrow grey stripes and the hind wing has a pale grey fringe. The whole life cycle is completed in about 14-58 days depending on weather conditions. There are several generations in a year.

Control

The pest can be controlled by repeatedly spraying the trees with methyl demeton / phosphamidon / dimethoate 0.03% at 15 days intervals. The collecting and burning the mined leaves is also useful to control this pest.

7. MANGO LEAF HOPPER

Idiocerus atkinsoni* Leth. (Homoptera: Cicadellidae) (Fig. 83)

It is a major pest of mango. The damage is caused by both nymphs and adults which suck the juice of the young shoots and flower buds of the entire inflorescence. Like aphids, the mango leaf hopper secretes honey dew which covers the branches of the trees and facilitates fungus growth.

The adult insect is small, wedge shaped with dark grey wings. The head is broad and prominent and possess three dark brown spots. The hind legs are thickly clothed with a number of bristles.

Life history

The female lays eggs singly inside tissues of the leaves or flower buds in a streak made by ovipositor. A single female lays 100-200 eggs which are dull white in colour. The eggs hatch in about 7-10 days releasing the nymphs which immediately ingest the juice from the

* Same as *Amritodus atkinsoni* Leth.

**Some workers have placed under Jassidae.

plant. They resemble the adult in all respects except in that they do not possess wings and have fairly long legs. There are usually 4-5 moults undergone by the nymphs in about 15-20 days and the entire life cycle is completed in about a month or so. There are several generations produced in summer while the adults hibernate in winter.

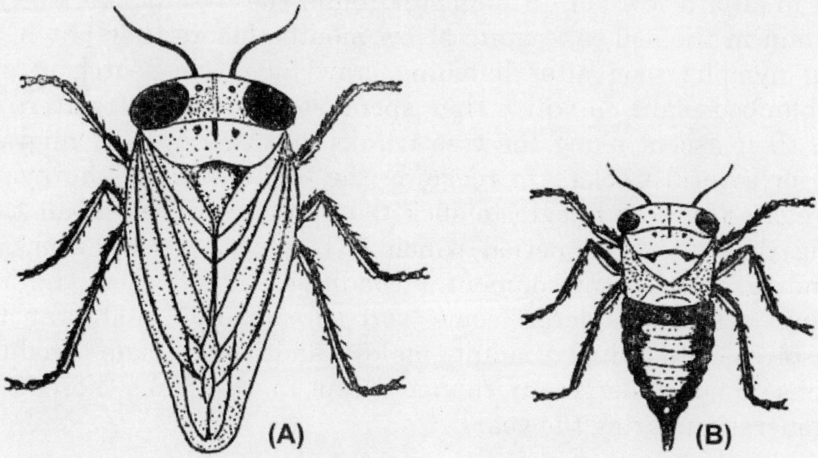

Fig. 83. Mango leaf hooper *Idiocerus atkinsoni*. **A.** Adult, **B.** Nymph

Control

Chemical control has been primarily employed to check the population of the mango leaf hoppers. This pest can be effectively controlled by spraying the trees with a number of insecticides. Some of the important being carbaryl 0.1% or diazinon or phosphamidon 0.02% or endosulfan or dimethoate 0.03% or malathion 0.05%.

8. MANGO MEALY BUG

Drosicha mangiferae **Green (Homoptera: Coccidae) (Fig. 84)**

These creatures have large fleshy flat bodies with a length about 1.5 cm. and breadth of about 1 cm. covered with ashy-white mealy powders and crawling up or down the tree-trunks or on the ground round the tree base or even invading the houses. Damage is caused by the nymphs only.

Life history

There is a well established sexual dimorphism in the adult stage which is generally found during the mid-summer period i.e. April - June. The adult females are wingless and large bodied. The male is a winged creature with only one pair of wings and a very delicate reddish body which flies actively. The male adults have much shorter longevity than the female adults which live for about one month. The adult gravid females, after fertilization, crawl down along the tree trunk to the ground where they lay eggs at the depths of about 5 -15 cm. and each cluster having 300 - 400 eggs. The oviposition is confined to an area of a few cm. in diameter round the base of the tree. The eggs laid in the soil take quite a few months before they hatch. The young nymphs soon after hatching crawl about in search of some suitable food plant on which they spend some time. Thereafter, they begin their ascent along the tree trunks and this upward migration lasts for several weeks. On reaching the fresh growths, the nymphs congregate there and begin to suck the plant sap. They moult thrice during their nymphal period which lasts about 3 months or more depending on the environmental conditions. Thereafter, the male forming nymphs undergo some sort of pupation and transform themselves into winged adult males, and the female producing nymphs do not undergo any change except in size. Thus there is only one generation during the year.

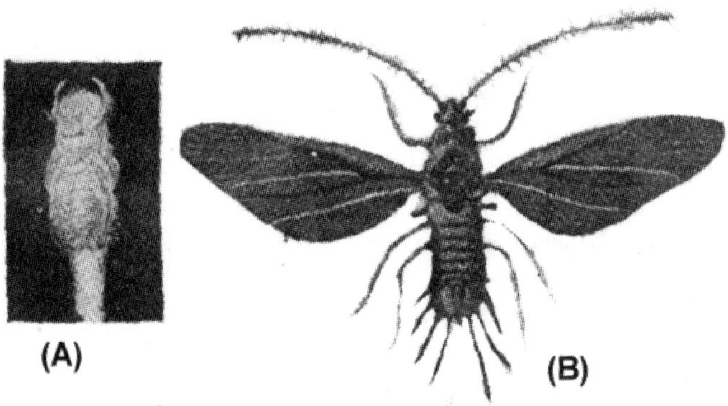

Fig. 84. Mango mealy bug *Drosicha mangiferae* - **A.** Male pupa, **B.** Male adult (From Lefroy, 1909)

Control

The pest can be controlled by destroying the eggs laid under the infested trees and application of sticky band round the tree trunk so as to check the nymphs from crawling up the trees. Incorporation of insecticides in the sticky band is likely to increase the effectiveness of this band. Even making a fairly wide insecticidal barrier round the trunk can lead to effective poisoning of the nymphs. The nymphs congregating below the bands may be killed by spraying them with methyl parathion 0.1%, diazinon or monocrotophos 0.04% or by burning or by shaking them off into a bucket of kerosinized water.

9. MANGO FRUIT FLY

******Dacus dorsalis*** **Hendel (Diptera: Tephritidae)**

The mango fruit fly is the most serious pest of all the fruit flies. It is widely distributed in India and other countries. Damage is caused by the grubs which feed on the pulp, thus rendering fruit unfit for human consumption. The apodous full grown maggots measure 8 - 9 mm. long and 1.5 mm. across the posterior end. They are yellow and opaque in colour. The adult is bigger than the housefly measuring about 14 mm. across the wings and 7 mm. in body length. It is brown, has transparent wings and, yellow legs with dark rust red and black patterns on the thorax.

Life history

The flies mature in about 15 days and lay 4-15 eggs at one time in clusters, deep into the soft skin of the mango fruit with the help of sharp ovipositors. Each female fly lays on an average 50 eggs but in favourable conditions 150-200 eggs are laid in one month. They hatch in 2-3 days in March-April, 1-2 days in the summer and in 10 days in the winter. The maggots pass through 3 instars in the ripening pulp and are full grown in about 6-29 days. Thereafter they leave the fruit, move away by jumping in little hops. They pupate inside the soil below 8-14 cm. The flies emerge in about 7-44 days. The whole life cycle is completed in 2-13 weeks depending on the climatic conditions and there are several generations in a year.

Control

The pest can be controlled by spraying the plants frequently with a mixture containing 0.05% malathion and 1% molasses or dimethoate 0.03% upto two weeks before the fruits are picked.

10. MANGO STONE WEEVIL

*******Sternochaetus mangiferae*** **Fabr. (Coleoptera : Curculionidae) (Fig. 85)**

The adult weevil is about 8 mm. in length and 4 mm. in breadth, greyish brown in colour and with such a pattern as to make the weevil

* Same as *Strumeta dorsalis* Hend.

** Same as *Cryptorrhynchus mangiferae* Fabr.

fit in the general back ground of the bark of the mango tree. In the non-fruiting season, this beetle remains practically in hiding under the bark and in other niches.

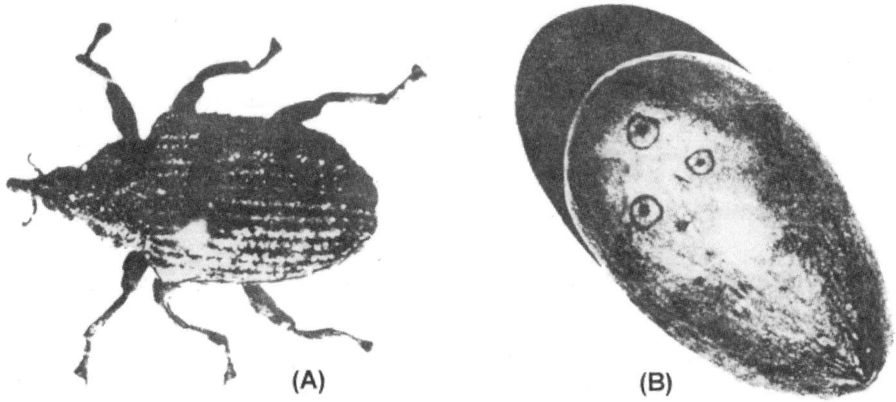

Fig. 85. Mango stone weevil *Sternochaetus mangiferae* Fabr. **A.** Adult, **B.** Mango fruit with three characteristic spots (enclosed in separate circles drawn on the mango surface) where the female weevil deposited its eggs and covered them with some secretion. (From Pradhan, 1969).

Life history

When the fruiting season arrives and the mango fruits are still quite tender, this beetle lays eggs on the surface of the mango fruit. The wound caused by the ovipositor heals soon after and the fruit does not exhibit any outward sign of infestation. One weevil lays about 15 eggs in a day and about 300 during a period of about 3 months and up to 3 dozen eggs can be met within a singly fruit. These eggs hatch within a week or so and the just hatched larva is cylindrical in shape. Inside the seed endosperm the larva feeds and moults 5 times during a period of about 5 weeks, pupates for about 7 days and finally transforms itself into the adult weevil. The generation is completed in about 40-50 days but the emerging adults become inactive and resume breeding in the next season only and thus, there is only one generation in a year. It is also interesting to note that the pest is monophagous and attacks only mango.

Control

The pest can be suppressed by destroying all fallen fruits, weeviled mangoes and by disposement of refuse stores and debris etc. The insecticidal control is not possible because the weevil being an internal feeder throughout its development.

11. MANGO STEM BORER

Batocera rufomaculata **Dejean (Coleoptera : Cerambycidae) (Fig. 86)**

It is a borer of a very large number of trees including mango. Beetle is about 4 to 6 cm. long, dark with a fine greyish vestiture, pronotum with 2 kidney shaped orange yellow spots, scutellum white elytra in basal third with numerous black tubercles and several yellowish spots variable in number and shape.

Life history

Each feamle lays about 200 eggs placed one by one in incisions cut in the bark with the mandibles. Oviposition normally takes place on dead trees but also on the branches and trunks of living trees that are not in good health and on the roots of trees exposed to erosion. Each egg is brownish white cylindrical 6 x 2 mm. long with narrow rounded ends. Newly hatched larva is about 10 mm. long. it feeds at first in the meristem and later penetrate deeper. The tunnels in the depth of the wood are irregular and 2 or more centimetres in cross section. The pupal chamber is 4-6 cm. cavity in a widened extension of part of the larval gallery and is surrounded by long interlaced fibres. The larva is bright yellow without any food in alimentary canal. The pupal period lasts about 3 to 4 weeks and is followed by an immature beetle stage of variable duration. The beetle emerges by a short tunnel running direct to the outside and ending in a circular exit hole.

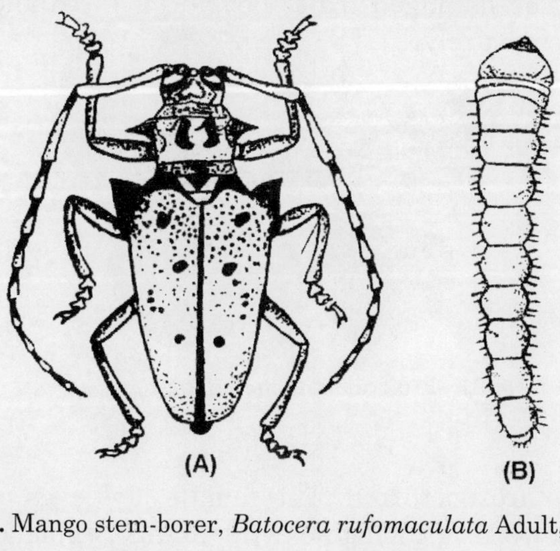

Fig. 86. A. Mango stem-borer, *Batocera rufomaculata* Adult, **B.** Grub.

The emergence period is from March to August, about 50% beetles emerges in May and 30% in June. The life cycle is thus annual.

Control

Removal of bark from logs shortly after felling prevents egg laying by all Cerambycidae. This is one of the standard measure for protecting sapwood and heartwood.

The damage to cultivated mango and other trees by *Batocera rufomaculata* can be prevented the stem with stout paper, coated with coal-tar or with wire gauge 1.5 mm. mesh or spraying bark with a strong repellent like Bordeaux mixture during the oviposition period.

12. FRUIT SUCKING MOTHS

Ophideres species (Lepidoptera: Noctuidae) (Fig. 87)

The fruit sucking moths are minor pests of mango, grapes, apple and citrus. They are distributed throughout India. The common species are *Ophideres conjuncta* Cramer, *O. materna* Cramer, *O. ancilla* Cramer and *O. fullonica* Linnaeus. The general body of *Ophideres conjuncta* Cramer is faint orange brown. Its forewings are dark grey and the hind wings are orange red having two black curved patches. The larvae are semiloopers. A fully grown larva is about 50-60 mm. in length.

Generally, moths and butterflies damage the plant or plant products only during their larval stages but fruit sucking moths damage fruits in their adult moth stage. The fruit sucking moth with the help of its strong piercing mouth parts punctures the fruit for sucking juice. The damaged fruits develop a rot round the puncture and drop off prematurely.

Fig. 87. Fruit sucking moth *Ophideres species*

Life history

The moths are nocturnal. The female lays eggs on a number of wild plants and weeds growing near the mango, apple and citrus trees. Eggs are round translucent about 1 mm in diameter. They hatch in about 1-2 weeks and within 24 hours of emergance, the young larvae start feeding on the leaves of host plant. The larva passes through 5 instars in 4-5 weeks. The pupa is dark red or brown in colour. The pupal stage lasts for about 2 weeks. The moth on emergence fly to nearby orchards for feeding on fruit juice. The exact duration of the life of the moth is not knwon. Probably there are 2-3 generations in a year.

Control

Spraying trees with diazinon 0.02% is effective to control this pest. Poison baiting with 20 gm malathion 50% W.P. or 50 ml diazinon + 200 gm gur or molasses in 2 litres of water attracts and kill moths.

13. SAN JOSE SCALE

***Aspidiotus perniciosus* (Comstock) (Homoptera : Coccidae) (Fig. 88)**

The pest is world wide in distribution. Several hundred species of the fruits as well as ornamental plants are severely damaged by this pest. It mostly prefers plants of the family Rosaceae, such as apple, plum, pear, peach and other related species. The damage is caused by nymphs and the female scales which suck the sap from twigs, branches, and fruits. The infested fruits also have scaly appearance and each spot is surrounded by a scarlet or red area.

Life history

Each female may give birth to 200-400 nymphs. The new born nymphs crawl out of the parental scale and lead a free life for 12-24 hours. These also suck the cell sap and are full grown in about 40 days. The females that develop again start giving birth to young ones within the next 10-14 days. The gravid mothers live 50-55 days. The male nymph has an elliptical or oval scale and develops into a winged adult in 25-31 days. The life of the male adult is 24-32 hours only, during which it fertilizes the non-winged females. There are 4-overlapping generations in a year and the fifth generation nymphs over winter.

Control

The pest can be suppressed by spraying the trees with diazinon or methyl parathion 0.05% or methyl demeton 0.03% during summer. It can also be controlled by spraying the trees from November – March when there are no leaves with Diesal oil emulsion + Bordeaux mixture (diesel oil 27 lit. + copper sulphate 6 kg. + lime 1.5 kg. to be emulsified and diluted 5-6 times before spraying) or with 3% Esso spray oil emulsion.

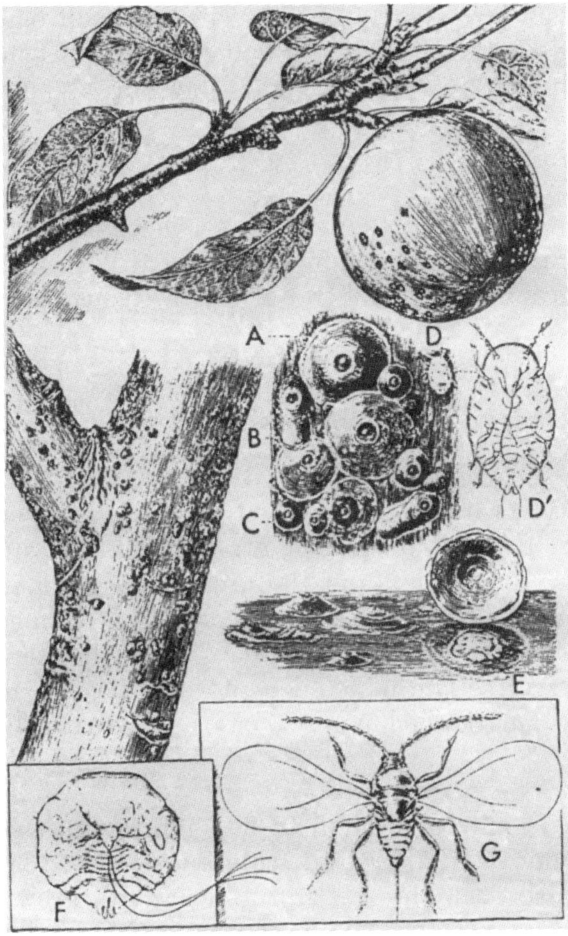

Fig. 88. San Jose scale *Aspidiotus perniciosus*, **A.** adult female scale, **B.** male scale, **C.** young scales, **D.** Nymph just hatched, **D'.** Same much enlarged, **E.** Scale raised, showing body of female beneath, **F.** Body of female much enlarged, **G.** Adult male much enlarged, (From Pfadt, 1962).

14. TENT CATERPILLAR

Malacosoma indicum Walk. (Lepidoptera : Lasiocampidae) (Fig. 89)

This is an important pest of apple and other fruits like pear, apricot and walnut. In serious infestation the whole plant may be defoliated and subsequently the caterpillars may feed even on the soft bark and twigs. When the infestation is severe 40-50% of the apple plants may be defoliated producing a poor harvest.

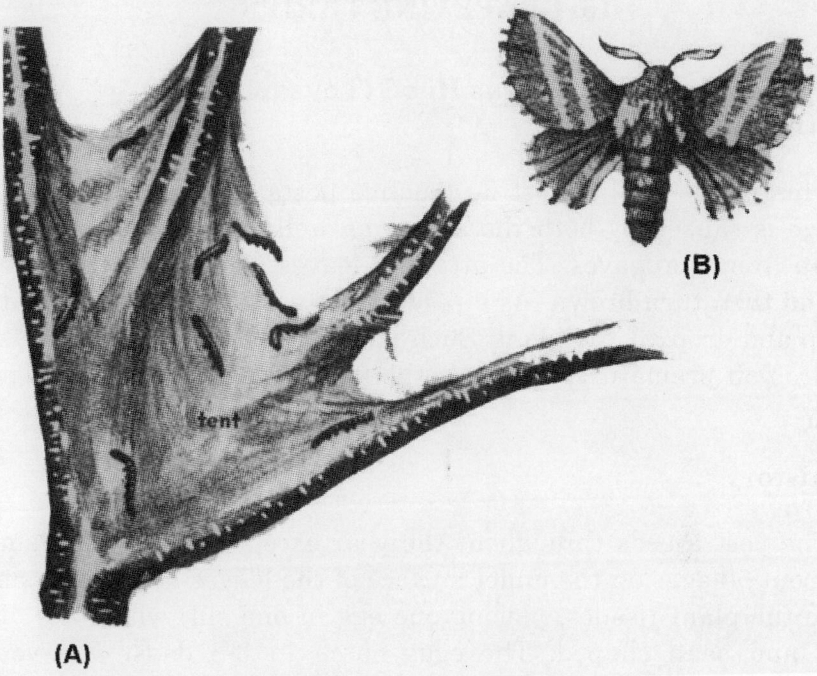

Fig. 89. Tent caterpillars *Malacosoma indicum*. **A.** Tent with caterpillars, **B.** Moth

Life history

The females lay eggs in broad bands around the branches. Each band may consists of 200-400 eggs. The larvae are gregarious. Soon after emergence each larva spins a silken nest on the tree. As the larvae grow the nest is also enlarged till it is 0.3-0.5 metre across. The full grown larva is 40-45 mm. long with its head and abdomen black. The larval stage lasts 39-68 days and when full fed, they spin oval white and compact cocoons, each about 25 mm. in length. The pupal stage lasts about 8-22 days. The moths emerge in about three weeks. The male moth is light reddish and the female moth is light brown in colour with a wing expanse of 29-32 mm. and 35-37 mm. in male and female respectively. The moths are short lived. The life cycle is completed in one year.

Control

The pest can be controlled by spraying the trees with malathion 0.5%. It can be suppressed by the destruction of egg clusters. Kerosinized water in an open vessel be placed below the tree so that the larvae that fall may be killed readily.

15. GRAPEVINE THRIPS

***Rhipiphorothrips cruentatus* Hood (Thysanoptera : Heliothripidae)**

This is one of the most destructive pests of grapevine in India. Damage is caused by both the adults as well as nymphs by sucking cell sap from the leaves. The attacked leaves show a whitish hue at first and then turn brown. As a result of attack, the leaves ultimately curl up and drop off the plant. Such vines either do not bear fruits or they drop off prematurely. Even if the fruits mature, they are of poor quality.

Life history

The pest breeds throughout the year except in winter. A female lays about 50 eggs on the under surface of the leaves by making small slits to the plant tissues, placing one egg in one slit; which are dirty white, and bean shaped. The eggs hatch in 3-8 days. The young nymphs appear reddish at first which become yellowish brown as they grow older. They feed on the lower surface of the leaves by rasping the surface and sucking the cell sap that oozes. They are full grown in 9-20 days and pupate on the leaves. In about 2-5 days they transform into adults which start feeding like nymphs. The adults are tiny about 1.4 mm. long, blackish brown with yellowish wings. The females can reproduce with or without fertilization. The fertilized eggs giving rise to females and unfertilized ones to males. The insect hibernate as "pupa" in the soil from December – March. There are several generations in a year.

Control

The pest can be controlled by spraying the vines with dimethoate 0.03%, when the vines are setting fruit, they should be sprayed with nicotine sulphate 0.04% or malathion 0.05%.

16. GRAPEVINE LEAFHOPPER

***Erythroneura species* (Homoptera : Cicadellidae) (Fig. 90)**

Numerous species of leaf hoppers, sucking the sap from the under surface of the leaves are found in the areas where grapes are grown. Damage is done by both adults and nymphs, which suck the sap and cause the foliage to become blotched with small white spots. When the infestation is heavy, the leaves turn pale or brown and fall from the vines. Due to severe infestation of leaf hoppers, the quantity and quality of the fruits are greatly affected.

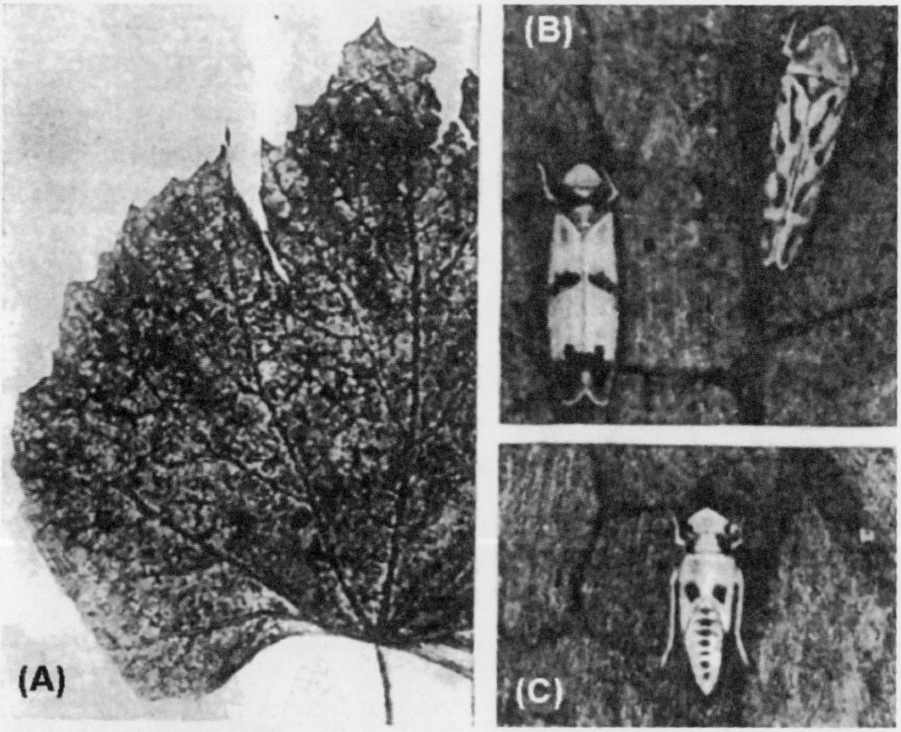

Fig. 90. Grape vine leafhoppers *Erythroneura species* and injury. **A.** White speckling of leaf due to feeding of leafhoppers; **B.** Three-banded leafhopper (left) and grape leafhopper (right); **C.** Nymphal leafhopper (From Pfadt, 1962)

Life history

Adults hibernate in winter season. During spring, they become active and feed on any green plant, before the grape foliage appears. Eggs are laid in the leaf tissues and they hatch in about 14 days. The yellowish wingless nymphs feed on the under surface of the leaves and moult 5 times before changing into adults. Developmental period requires 3-5 weeks depending on the temperature. There are 2-3 generations in a year.

Control

The leaf hoppers can be controlled by spraying malathion 0.05%, dimethoate 0.03% or endosulfan 0.03% or carbaryl 0.01%.

17. GRAPEVINE GIRDLER

Sthenias grisator Fabr. (Coleoptera : Cerambycidae) (Fig. 91)

Grapevine girdler occurs throughout the grape growing areas of the country. Besides grapes, it is also a pest of almond, cashew, mulberry, mango, jack fruit and several garden plants. The damage is caused by girdling of the adult beetles which is followed by the drying up of the wood. The adult is a medium sized, stout beetle. The beetle is greyish brown in colour and is provided with white and brown irregular markings. The insect resembles a mottled bark. The pest prefers girdling of the green branches before laying eggs, which causes a lot of damage to the grapevines as a result of which the vines dry up above the level of girdling.

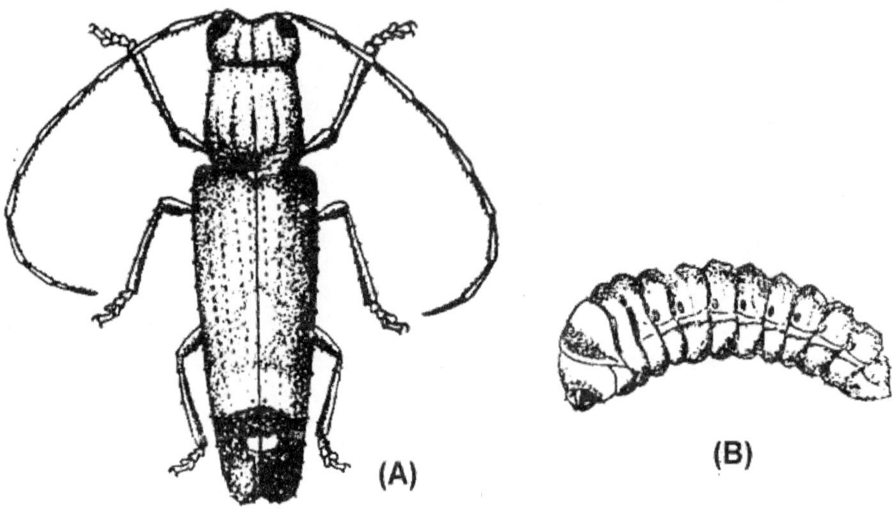

Fig. 91. Grapevine girdler *Sthenias grisator*. **A.** Adult, **B.** Larva

Life history

The eggs are deposited at night underneath the bark in cuts made by the beetle on girdled vines. Eggs are laid in groups of 2-5 which are oval in shape and measure about 3-4 mm. in length and 1 mm. in width. The eggs hatch in about 7-8 days. The freshly hatched grubs are very small measuring about 2-4 mm. in length and soon tunnel into the wood. The head is dark brown and the thorax is globular provided with chitinous spines. As with other cerambycids, the mouth parts, specially a pair of mandibles are very prominent and powerful. The full grown grub is about 10-12 mm. long, which pupates in the tunnel. The whole life cycle is completed in more than a year.

Control

Collection and destruction of the girdled branches is a simple method to control the pest. Hand picking adult beetles is also very effective. The pest can be controlled by spraying the grape vines at the ground level with malathion 0.05%.

18. GRAPEVINE LEAF ROLLER

***Sylepta lunalis* Guenee (Lepidoptera : Pyralididae)**

It is a serious pest of grapevine, particularly during the late monsoon season. It is distributed to all the grape growing parts of the country. The damage is done by the caterpillars. The newly hatched and young caterpillars feed on the green chlorophill of the leaves and skeletonize them. On the other hand the mature caterpillars roll up the leaf lamina. It is interesting to note that each leaf roll contains only one caterpillar. The adult moths are pale-brown or ochrous in colour and are provided with white spots on both the forewings as well as the hind wings. Males are smaller than the females e.g. the wing span of the male moth is 21-25 mm. whereas that of the female moth is 24-28 mm.

Life history

Each female moth lays about 100-120, creamy white, oval eggs on the under surface of the leaves. The eggs measure about 1.5 mm. Hatching take place in about 3-4 days. The newly emerged larva is very small pale-green in colour and is about 3-3.5 mm. in length. The mature caterpillars measure about 26 mm. in length and their body is covered with grey-brown hairs. The head is drak brown or black in colour. The larva passes through 5 instars and pupation takes place within the leaf rolls. Later on, the pupa drops down onto the fallen dry leaves and debries. The pupal stage lasts 5-8 days, after which the moth emerges. There is normally one generation in a year.

Control

The pest can be controlled by spraying on the vines with any one of these insecticides: malathion 0.05%, dimethoate 0.04%, phosphamidon 0.04%.

22

PESTS OF VEGETABLES

1. RED PUMPKIN BEETLE

*[*] *Aulacophora foveicollis* Lucas (Coleoptera: Chrysomelidae) (Fig. 92)**

It is a serious pest of cucurbitaceous plants like melon, pumpkin, cucumber etc. The attack caused by the feeding of adult insect or its larva on leaves, is indicated by the formation of irregular holes in the leaves. The larvae also damage the shoots, roots of the plants which ultimately withers and dies. The chief damage however is done to the fruits which are rendered useless for human consumption.

The adult is a brilliant orange coloured beetle, with the body surface shiny and smooth. The underside is entirely black and is covered over with short, wide, soft hair like setae.

Life history

The female lays about 60-300 orange or yellowish-brown eggs within a period of slightly more than a month, either singly or in clusters on the moist soil underneath the plants. These eggs under optimal humidity and temperature conditions, hatch in about 6-15 days into young larvae. Right from the start the larvae attack the roots of the plants and moult 4-times before changing into pupa. For every moult the larva goes into the soil and casts off its cuticle. After a larval period of 13-23 days, they penetrate into the soil to a depth of 1 cm. to 25 cm. where they construct an oval waterproof creamy white cocoon. The pupal period lasts for 7-17 days after which the adult escapes out of the soil. The entire life cycle ranges from 30-55 days. Normally there are 3-5 broods in a year. These pests are extremely

---

[*] Same as *Raphidopalpa foveicollis* (Lucas)

active from March to October although occasionally they are found damaging plants in winter months also.

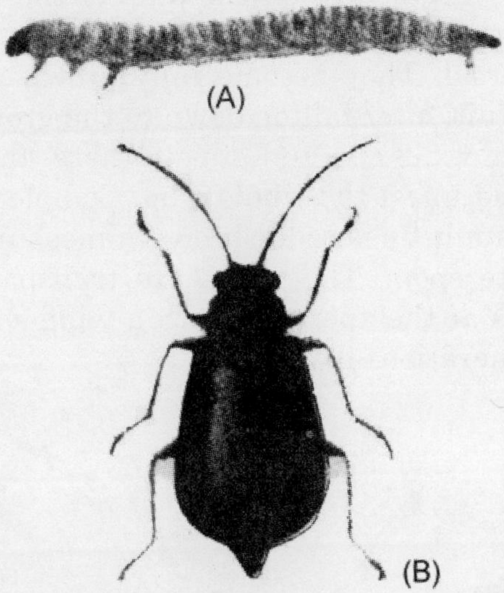

Fig. 92. Red pumpkin beetle *Aulacophora foveicollis* - **A.** Larva, **B.** Adult

Control

Besides certain precautionary measures such as adjustment in the time of sowing of plants, ploughing after harvest etc. it is the chemical control mostly which has been found to be effective. Using the carbaryl dust is very effective. Spraying the crop with dimethoate 0.02% or malathion 0.05% also controls the pest.

2. MILON FRUIT FLY

***Dacus cucurbitae Coqu. (Diptera: Tephritidae) (Fig. 93)**

This is the most destructive pest of melons and other cucurbits like ribbed gourd and bitter gourd, etc. Tomato is also attacked occasionally. It is distributed throughout India. The damage is caused by the maggots which bore into the fruits and feed on the pulp. The infested fruits can be easily recognized due to deformation or rotting area round the seat of oviposition.

Life history

The female fly lays 5-27 eggs singly or in batches underneath the rind (skin of young fruits) which is punctured by the ovipositor. The

* Same as *Strumeta cucurbitae* Coqu.

eggs hatch in 2-5 days liberating small dirty white apodous larvae. The larvae are typical dipterous maggots, 10-12 mm. when fully mature. These feed continuously on the pulp of the fruit resulting in its complete destruction. They can jump to a height of 30 cm. which helps them in dispersal. They become fully grown within 8-15 days and then leave the fruit; after falling down to the ground, they burrow in the soil to a depth of 2-8 cm. and pupate there. In about 5-10 days adult flies emerge and infest the whole crop. Complete life cycle takes about 3 weeks. The adult fly is reddish brown measuring about 5 mm. with black and white spots. The wings are transparent with brown bands and grey spots at the apex and with a wing-span of 12-15 mm. There are several generations in a year.

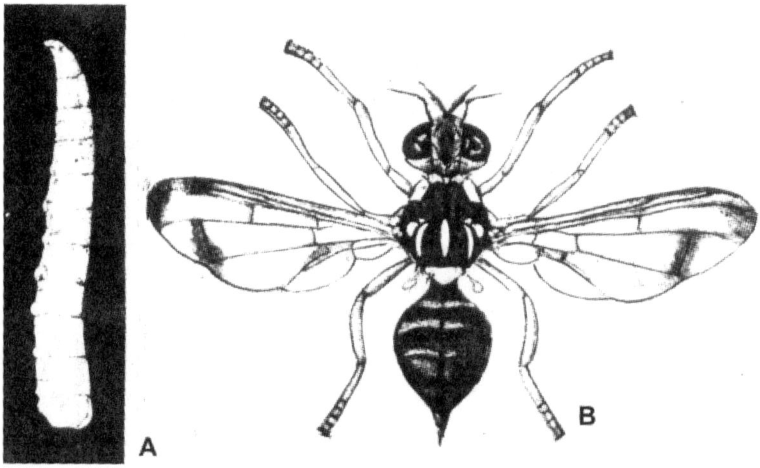

Fig. 93. Million fruit fly, *Dacus cucurbitae* **A.** Larva, **B.** Adult (From Fletcher, 1914)

Control

All infested fruits should be collected and destroyed as soon as they have fallen to kill the maturing maggots inside. In addition, use of 0.03% dimethoate, 0.05% malathion and 0.15% trichlorphon in a sugar solution form of spray, on the foliage, is also very effective.

A poison bait (20 gm malathion 50% W.P. or 50 ml of diazinon + 200 gm of gur or molasses in 2 litres of water) kept in flat containers attracts and kill flies.

3. DIAMOND BACK MOTH

Plutella xylostella (Linn.) (Lepidoptera : Plutellidae) (Fig. 94)

This insect is world wide in distribution and is a serious pest of cauliflower and cabbage. It also feeds on other cruciferous, solanaceous and leliaceous plants. The damage is cased by the caterpillars which in earlier stages feed in mines on the under side of the cabbage leaves and in the later stages feed on exposed leaves.

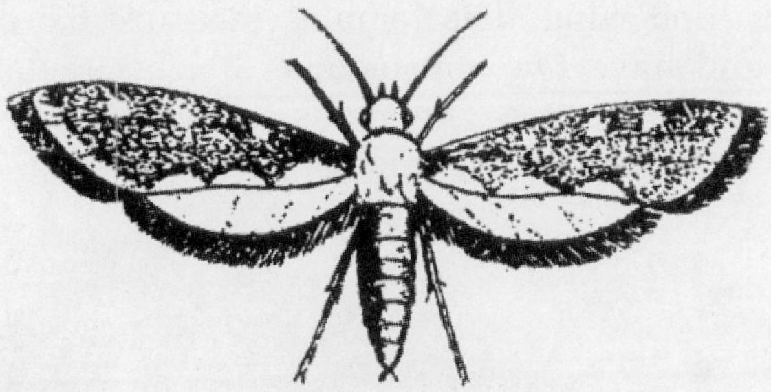

Fig. 94. Diamond-back moth, *Plutella xylostella*

Life history

The diamond-back moth is active throughout the year. The tiny yellowish eggs are laid singly or in clusters of 2-40 on the lower side of the leaves. A female lays about 18-356 eggs in her life time. The eggs hatch in about 3-10 days depending on the climatic conditions. They mature in about 9-17 days under different temperatures and measure about 8 mm. in length. They are yellowish green in colour with black hairs distributed all over the body.

Before pupation, the larva constructs a barrel shaped silken cocoon which is open at both the ends. The pupal stage lasts 4-5 days. The moths measure about 8-12 mm. in length, brown or grey in colour with distinct white spots on the fore wings and appear like diamond patterns while the wings lie flat over the body. The moths may live for about 20 days. The life cycles is completed in about 15-18 days during September-October and there are several generations in a year.

Control

The pest can be controlled by spraying the crop repeatedly at weekly intervals with a number of insecticides viz. malathion 0.05% or endosulfan 0.05% or carbaryl 0.15% etc.

4. CABBAGE BUTTERFLY

***Pieris brassicae* Linn. (Lepidoptera: Pieridae) (Fig. 95)**

The cabbage butterfly is a serious pest of almost all the vegetable plants of the family Cruciferae. It is world wide in distribution. The damage is caused only by the larvae which are known as "cabbage caterpillars". The butterflies are pale white with a smoky shade on the dorsal side of the body. The wings are pale white with a black patch on the apical angle of each fore wing and a black spot on the costal margin of each hind wing. The females measure 6.5 cm. across the spread wings and have two conspicuous black circular dots on the dorsal side of each fore wing. The males are smaller than the females and have black spot on the under side of each fore wing.

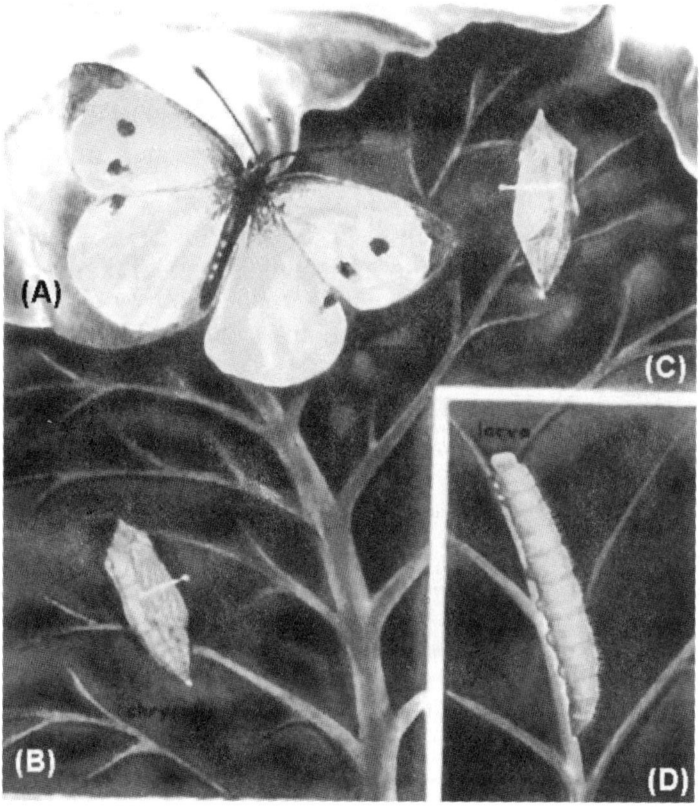

Fig. 95. Cabbage butterfly *Pieris brassicae* **A.** female butterfly, **B-C.** Chrysalis (Pupa) **D.** Larva

Life history

The butterflies are very active in the field and lay about 164 yellowish conical eggs in clusters of 50-90 on the upper or lower side of a leaf. The eggs hatch in about 11 - 17 days in November-February and 3 - 7 days in March - May. The caterpillars feed gregariously during early instars. The mature caterpillars measure 40 - 50 mm. in length. The larvae are pale yellow when young and turn greenish yellow later on. The head is black and the whole body is marked with black spots. They pass through 5 stages and are full fed in 15 - 22 days during March - April and in 30 - 40 days during November - February. Then they pupate in barns or on the trees. The pupal stage lasts 8 - 15 days in March - April and 20 - 28 days in November - February. The butterflies live for 3 - 13 days and breeds 4 times in a year.

Control

The pest can be controlled by spraying the crop with malathion 0.05%, diazinon 0.02% up to three weeks, before harvest. During the later stages of growth of the plants 0.2% carbaryl is very useful and safe.

5. CABBAGE SEMILOOPER

Plusia orichalcea (Fabr.) (Lepidoptera: Noctuidae) (Fig. 96A,B)

The cabbage semi-looper is widely distributed in India. It is a minor pest of cabbage, cauliflower and other winter vegetables. The semiloopers are yellowish-green in colour and are polyphagous. The damage is caused by the semiloopers, which by biting mouth parts, make circular or irregular holes of different sizes into the leaves of cabbage, cauliflower and other vegetables of the winter season. The adults are yellowish brown, with a large golden patch on each forewing. They measure about 42 mm. across the spread wings.

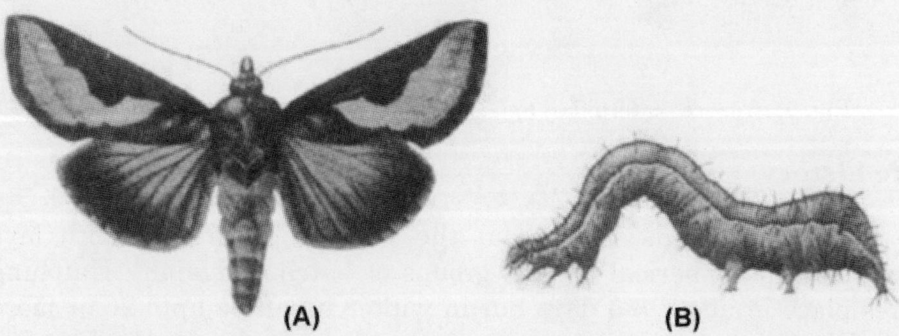

Fig. 96. Cabbage semilooper *Plusia orichalcea*. **A.** Adult, **B.** Semilooper

Life history

The cabbage semiloopers are active during winter (November-March). During active period they lay eggs on the leaves of cabbage, cauliflower and other cruciferous plants. The semiloopers feed individually, making holes on the leaves, the size of which varies considerably depending on the larval instar. The full grown larvae pupate in the debris or rubbish lying on the ground. Similar to many other moths, the moths of cabbage semiloopers are also very active at dusk i.e. crepuscular.

Control

The pest can be controlled by spraying the crop with malathion 0.05% or diazinon 0.02% upto three weeks before harvest.

6. TOBACCO CATERPILLAR

**Spodoptera litura* (Fabr.) (Lepidoptera : Noctuidae) (Fig. 97)

The tobacco caterpillar is world wide in distribution. In addition to tobacco, it feeds on castor, groundnut, tomato, cabbage, pulses, solanaceous and cruciferous crops. The damage is caused by the caterpillars which feed on leaves and are nocturnal.

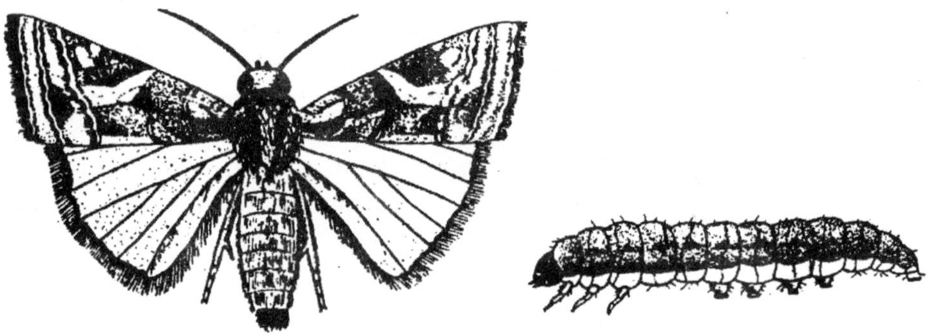

(A) (B)

Fig. 97. Tobacco caterpillar, *Spodoptera litura* **A.** Adult, **B.** Caterpillar.

Life history

The pest breeds throughout the year. Each female moth lays about 100-300 spherical eggs in groups or batches at night. Hatching takes place in about 3-5 days but in winter it can be upto 26 or more days. The newly hatched larvae are gregarious feeder for the first few days but later they disperse and feed individually. They pass through 6 stages and are full fed in about 15-30 days. The mature larvae measure 35-40 mm. in length. They are velvety black with yellowish green colour on the dorsal side and white bands on the lateral side. Pupation takes place in the soil in an earthen cell or under fallen

* Same as *Prodenia litura* (Fabr.)

leaves. The pupal stage lasts about 7-15 days. The moths are about 22 mm. long and measure 40 mm. across the spread wings. The fore wings have golden and greyish brown patterns. The hind wings are white and margins of the wings are brown. The moths after emergence live for 7-10 days. The entire life cycle is completed in about 35-60 days depending on the climatic conditions. There are about 8 generations in a year.

Control

Cultural methods of control such as ploughing and burning of crop stubbles, flooding the infested fields and removal of weeds, all helps to lower the pest population. The pest can be controlled by spraying the crop with diazinon 0.02% or phosphamidon 0.03% or endosulfan 0.05%. Dusting the crop with carbaryl is also very effective. The infested crop should be sprayed with malathion 0.05% or carbaryl 0.15%.

7. EPILACHNA BEETLES

***Epilachna species.* (Coleoptera: Coccinellidae) (Figs. 98, 99)**

The *Epilachna* beetles mostly attack different solanaceous and cucurbitaceous vegetables, as well as some other crops. Their feeding habit is quite characteristic and both the adult and larval stages skeletonize the leaves which present a lace like appearance as green matter in between the veins is eaten away leaving the skeleton of anastomosing veins.

Life history

The eggs are laid in clusters, varying in number, usually on the underside of the leaves and placed vertically. The eggs are pale yellow, elongate or oval with comb like hexagonal sculpturing and they measure about 0.5 mm. in length. Incubation takes about 4-5 days. The larvae are pale yellow, covered with delicate spines when first hatched. The young larvae start feeding soon after hatching making rows of small windows in the leaves. Full grown larvae are dark yellow, broad with a dark head and strong branched spines and 6-7 mm. long. Larval development takes about 16 days. Pupation takes place on the leaves of the host plant and the pupa is dark yellow. The adult beetles are oval, 6-9 mm. long, reddish to brownish yellow but colour is variable. Each elytra is marked with a series of black spots. *Epilachna* beetles are the only phytophagous representatives of the family Coccinellidae. The whole life cycle takes about 30-35 days under suitable conditions.

Fig. 98. Life stages of *Epilachna* beetle (showing eggs, larva, pupa and adults)

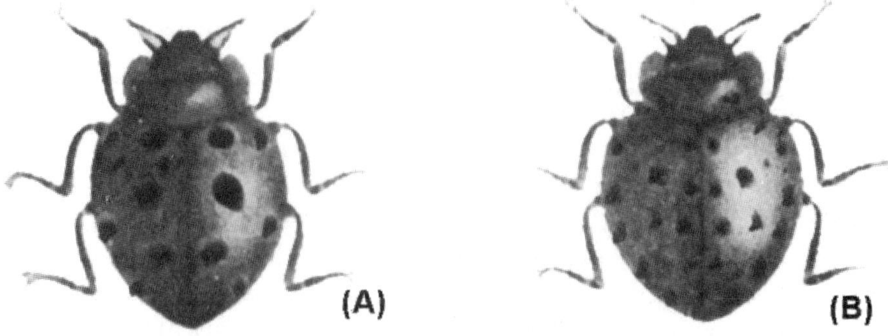

Fig. 99. Two common *Epilachna* beetles: **A.** *Epilachna dodecastigma;* **B.** *Epilachna* 28-punctata (From Fletcher, 1914)

Control

The *Epilachna* beetles can be controlled by spraying them with diazinon 0.02% or malathion 0.05% or 0.1% carbaryl. On cucurbitaceous vegetables, only malathion 0.05% or methyl parathion 0.02% should be sprayed as the other insecticides are phytotoxic.

8. BRINJAL FRUIT BORER

Leucinodes orbonalis Guen. (Lepidoptera: Pyraustidae) (Figs. 100, 101)

This is commonly known as "brinjal fruit borer". The pest is widely distributed in several countries. The young caterpillars are creamy white, the mature larvae are pinkish in colour. The moths are white having brown or black spots on the dorsal side of the thorax as well as abdomen. The wing margins are provided with fine bristles and the fore wings have numerous black or brown spots. The damage is caused mainly by the caterpillars which attack the terminal part of the shoots, thus destroying the growing point. Fruit-buds and fruits are also badly damaged by borers specially in autumn. Further, different varieties of brinjal have been reported to show differential susceptibility to this pest. According to some workers, the varieties of long, narrow fruits to be less infested than those with spherical fruits.

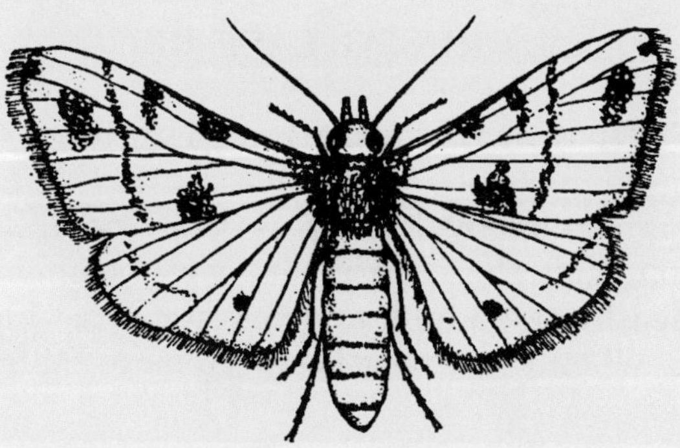

Fig. 100. Brinjal fruit-borer, *Leucinodes orbonalis,* moth

Life history

Moths of the brinjal borer appear in the beginning of summer i.e. in late March. Each female lays about 100 or 120 eggs singly or in groups of 3 to 4 on almost all the parts of the plant i.e. leaves, stem, flowers and tender fruits. Hatching occurs in about 4 to 6 days. The newly emerged larvae bore into the terminal part of the shoot, flowers, fruits etc. where the eggs are laid. There are 5 instars and the larval stage lasts 1 to 4 weeks. The mature larvae pupate in silken cocoons among the fallen dry leaves. Pupal duration is about 1 to 2 weeks. The whole life cycle is completed in 3 to 6 weeks. There are several overlapping generations in a year.

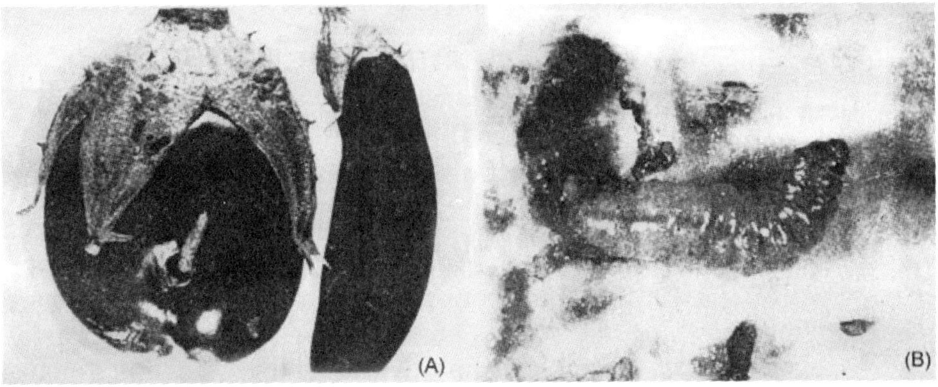

Fig. 101. A. Brinjal damage by brinjal fruit-borer, **B.** Larva of the brinjal fruit-borer. (From Atwal, 1976)

Control

The borers can be controlled by spraying, malathion, 0.04% or diazinon 0.02% or 0.1% carbaryl. The borers can be destroyed by removing the infected fruits and leaves.

9. BRINJAL LACE BUG

Urentius sentis Distant (Heteroptera : Tingidae) (Fig. 102A)

This bug is commonly known as the brinjal lace bug and is distributed throughout the country but very common in plains. The damage is caused by the adults and the nymphs which suck the sap from the leaves. When the attack is severe more than 50% of the crop may be destroyed.

Life history

The bugs live for about 30-40 days and lay 35-44 shining white nipple-shaped eggs singly in the tissues on the underside of the leaves. The eggs hatch in 3-12 days and the nymphs feed gregariously on both the lower as well as upper surface of the leaves. They grow through 5 stages and transform into adults in about 10-23 days. The insect passes through 8- overlapping generations in a year.

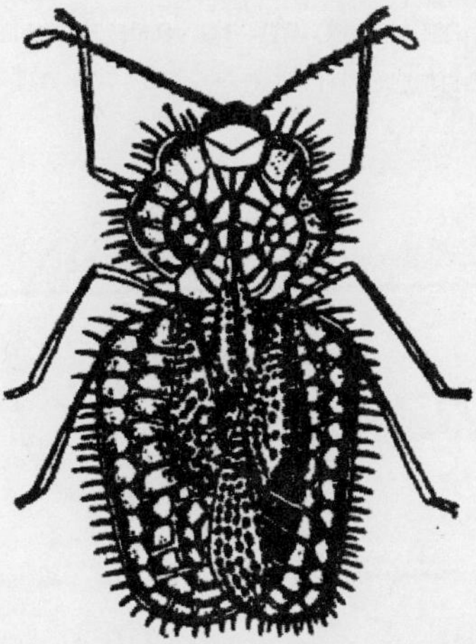

Fig. 102 A. Brinjal lace bug, *Urentius sentis,* adult

Control

The pest can be controlled by spraying the crop with diazinon 0.02% or dimethoate 0.05%.

10. POTATO TUBER MOTH

**Gnorimoschema operculella* (Zeller) (Lepidoptera : Gelechiidae) (Fig. 102 B)

It is a destructive pest of potato, tobacco, tomato, brinjal, and several other vegetable plants. The damage is caused only by the larvae, which are pinkish white or greenish in colour. The head of the caterpillar is dark brown.

Life history

In favourable conditions the insect may breed throughout the year. As with several other insects, in cold weather the life cycle is much prolonged and in warm and moist weather, it is much shorter. A female moth lays about 150-200 eggs, singly on the under side of the leaves as well as on the exposed potato tubers. The caterpillars first produce blotch mines on the leaves and then they work their way into the stems. The larvae are about 20 mm. in length. The larval stage in summer lasts 2-3 weeks and the mature larvae pupates in a greyish silken cocoon which is about 13 mm. in length. The moths emerge in 7-10 days and are very small in size about 13 mm. across the wings

* Same as *Phthorimaea operculella* (Zeller)

when spread. It is greyish brown in appearance. The life cycle is completed in about a month and there are about 5 or 6 generations in a year.

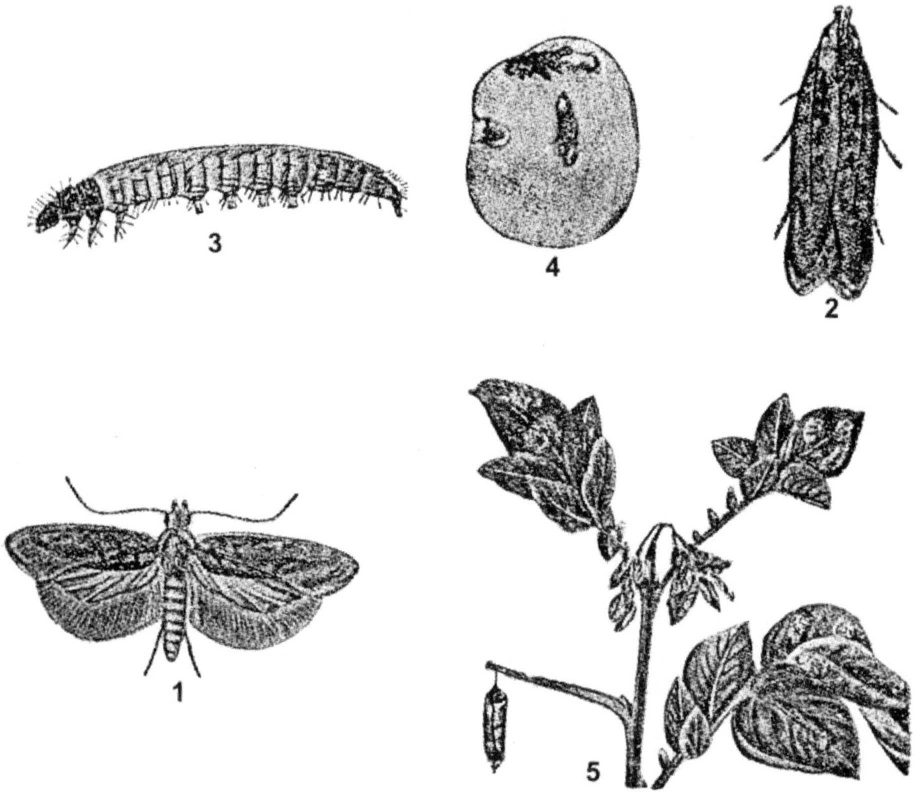

Fig. 102B. Potato tuber moth, *Gnorimoschema operculella*
1. Female moth, **2.** Male moth, **3.** Full grown caterpillar, **4.** Caterpillar inside infested potato tuber. **5.** Part of infested potato plant. (From Pruthi, 1969).

Control

Dusting the potatoes meant for seed with malathion 2% is very effective to control the pest. If potato tubers in the stores are infested, fumigation with carbon disulphide @2-3 litres/100 cubic meters is recommended to control this pest.

23

PESTS OF PLANTATION CROPS

The plantation crops viz. tea (*Camellia spp.*), coffee (*Caffea spp.*) coconut-palm (*Cocos nucifera*) etc. are very important and useful from economic point of view. They are perennial and provide ideal situation for intensive pest infestation. There are a large number of insect pests of plantation crops. In the following pages only a brief account of some of the most destructive insect pests of tea, coffee and coconut palm plantation crops, is given.

PESTS OF TEA

1. BUNCH CATERPILLAR

***Andraca bipunctata* Walk. (Lepidoptera: Notodontidae) (Fig. 103)**

The larvae are known as "bunch caterpillars" because they are gregarious in nature. The moths measure about 40 to 50 mm. across the wings. They are brown in colour. Besides the wavy cross lines, the forewings have two white spots near to the outer margin. The hind wings are rather pale in the anterior region and brown in the posterior region.

Life history

Each female moth lays about 500 yellowish eggs in linear clusters on the underside of the tea leaves. Hatching of eggs takes place in about 7 to 10 days. The just hatched larvae are yellowish in colour. They eat up their egg-shell before starting feeding on leaf tissues. There are 5 instars and the total time taken for development varies from 3 to 4 weeks. The fully mature larva constructs a cocoon in between the dry leaves of the tea. The full grown larvae are 50 to 65 mm in length. The pupa is about 25 mm in size and reddish brown in colour. The pupal period lasts 2 to 4 weeks. There are generally 4 generations in a year in the north eastern region of India. The larvae feed on the tea leaves thus causing a considerable loss to the tea crop.

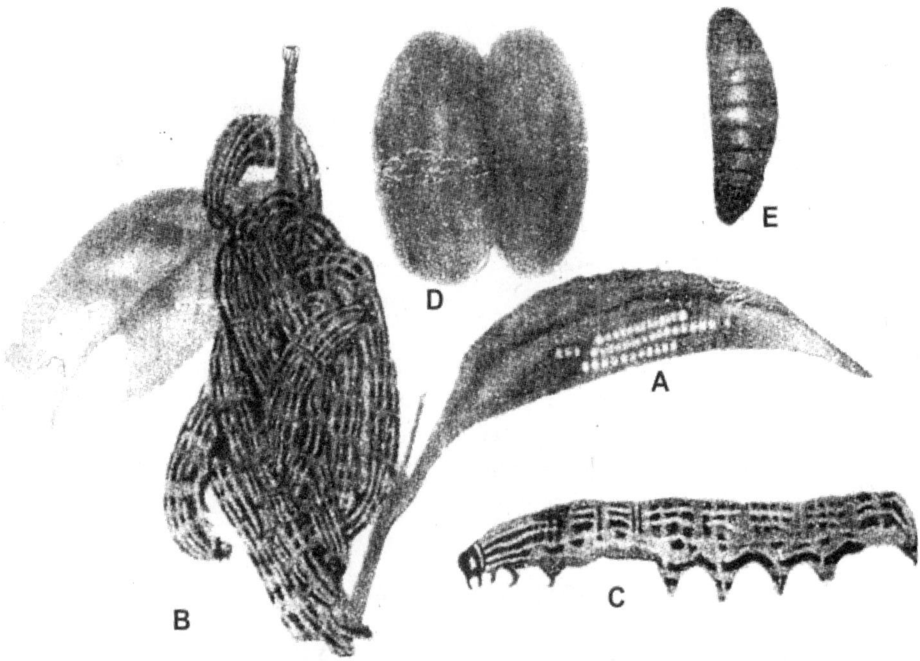

Fig. 103. Bunch caterpillar, *Andraca bipunctata* **A.** Egg clusters, **B.** Cluster of caterpillars, **C.** Caterpillar, **D.** Cocoon, **E.** Pupa (From Pruthi, 1969).

Control

In addition to hand picking and destroying, insecticidal methods are usually used for the control of tea pests. Spraying of 0.05% malathion or 0.05% endosulfan or 0.02% carbaryl was found useful to kill the larvae.

2. TEA MOSQUITO BUG

Helopeltis theivora Wat. (Heteroptera: Miridae) (Fig. 104)

This is commonly known as "Tea mosquito bug". The female bug is orange and the male bug is black in colour. Since the bugs are provided with piercing and sucking type of mouth parts, they cause considerable damage in nymphal as well as adult stages. The nymphs and adults both suck the sap from leaves, buds and delicate parts of the tea stems. There are two species of this bug, the *H. antonii* is found in South India and the *H. theivora* in eastern part of India.

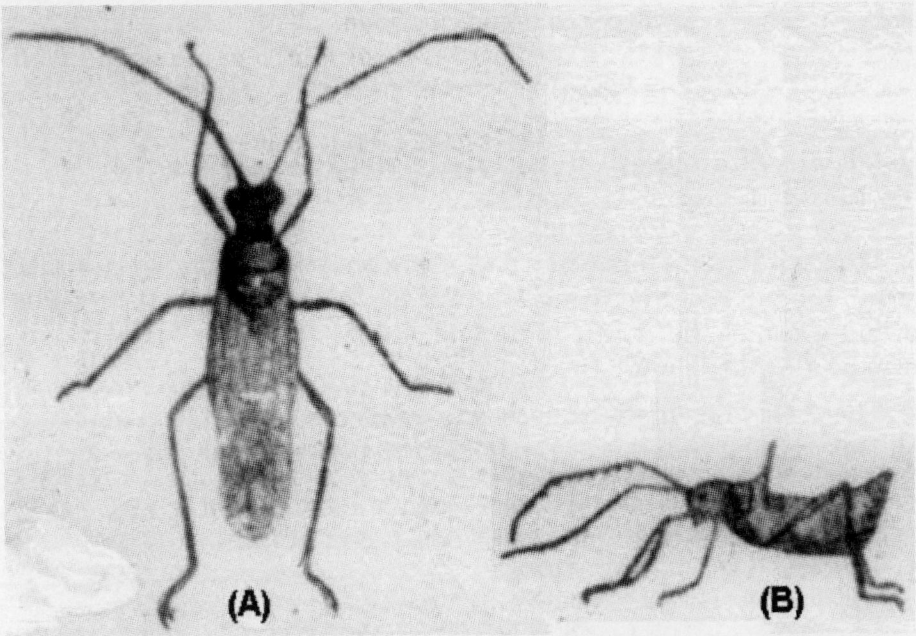

Fig. 104. Tea mosquito bug, *Helopeltis theivora* Waterh. **A.** Adult, **B.** Nymph.

The damage is caused to the young tea leaves by making punctures and sucking the sap. When large number of punctures are made, the leaves of the tea appear black and fall off.

Life History

About 500 eggs are laid by each female bug on the buds of tea. The eggs are like rice-grains which hatch in about 5 to 27 days depending on the prevailing temperature. The nymphs have 5 instars and the bugs complete their development in 2 weeks in warm season. In cold climatic condition, complete development of the nymph from 1st to 5th instar takes about 8 to 10 weeks. The adult bug is slender, small measuring about 6 to 8 mm. in length. They are nocturnal and remain hidden under the tea leaves during the day time. This bug, as with other bugs, prefers humid and warm climate. The tea bushes provide such suitable conditions for this bug.

Control

The incidence of the pest can be suppressed by collecting and destroying the nymphs as well as adults with hand nets in the morning and evening. The use of 0.05% malathion, as in other cases has also been found very effective to kill the bugs.

PESTS OF COFFEE

1. COFFEE STEM BORER

Xylotrechus quadripes Chev. (Coleoptera: Cerambycidae) (Fig. 105)

It is the larval stage that bores into the main stem some times down to the tap root and primary branches. It is common in the coffee growing areas of the South India and Assam. The adult is blackish-brown beetle a little more than one cm in length.

The main damage is caused by the larvae which damage coffee plants by making tunnels as a result of which the branches break.

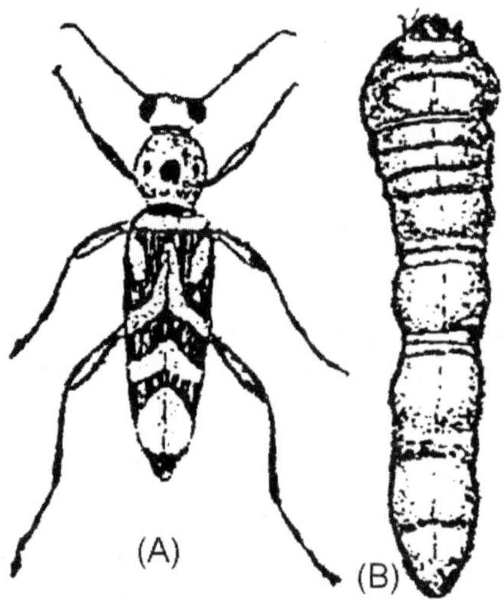

Fig. 105. Coffee stem borer - *Xylotrechus quadripes* **A.** beetle, **B.** borer (grub)

Life History

A female beetle lays as many as 100 eggs in about 4 weeks time. The larvae which are white or grey-white in colour, hatch in about 10 days and start feeding as well as boring the stem of the coffee. The larvae feed and make tunnels in the stem. The fully developed larvae pupate inside the tunnels. The pupal period lasts about a month after which the adult beetle emerges and starts the next generation. There is only one generation in a year.

Control

The larvae can be picked and destroyed and the infected parts of the plants can be removed to prevent further damage.

2. COFFEE SHOT HOLE BORER

***Xylosandrus compactus* Eichh. (Coleoptera: Scolytidae)**

The beetle in appearance is brown in colour and cylindrical in shape. It has been found that the larvae and adults both form small holes in the bark hence the name "coffee shot-hole borers." The lower side of the plant parts i.e. stems and bark is preferred. The attack of this beetle is seen by the wilting, which is due to large number of holes. The damage is caused by the larvae and adults both by forming tunnels which are responsible for wilting of the coffee plants as well as falling of the leaves. The affected stems are completely damaged by the attack of micro-organisms.

Life history

The eggs are laid in tunnels of old braches. The number of eggs laid by a female individual is up to 50 or more. The eggs hatch in about 7 days under favourable condition. The larvae after hatching inside the tunnel, feed on various fungi growing in the tunnels on the faeces of the beetles. The larval period lasts about 3 weeks. The cocoons are made by the full grown larvae near to the exit hole of the future adult. The pupal development is completed in about 10 days. Thus, the total time taken to complete the life cycle varies from 35 to 40 days. Depending on the suitable conditions, there are several generations in one year.

Control

Hand picking and destroying the larvae and adults is the simplest method of control of this pest. Use of certain chemicals in the form of spray like malathion - 0.05% kills the insects.

PESTS OF COCONUT PALM

1. RHINOCEROS BEETLE

***Oryctes rhinoceros* Linn. (Coleoptera: Scarabaeidae) (Fig. 106)**

The adult beetle is known as "Rhinoceros beetle" because it has a pointed horn on the head. The body of the beetle is robust and elongate measuring about 5 cm. Unlike other beetles, the wings are well developed and it is a good flier. The adult beetle is one of the most destructive pests which feeds on the crown of the coconut trees. The larvae are harmless.

Life history

Each female beetle lays about 100 to 150 oval eggs, white in colour, which are laid 10-15 cm. below the soil level in decaying vegetable matter viz. decaying leaves, logs and stem of palms, rotten vegetables, old cattle dung, wet saw dust etc. The larvae emerge in about 10 to 15 days. The larval development takes about 100 to 180 days and passes through 3-instars. The full grown larva has an average length about 75 mm. The body is covered with short brownish hairs. Pupation takes place in the earthen chamber. The pupa is brown in colour and its size varies from 50 to 70 mm. The pupal period lasts for 2 to 4 weeks and even after emergence from the pupal shell the beetle remains in the cocoon for about 11-20 days before coming out of the soil. The adult beetle emerges in about 10 to 25 days. The beetles are nocturnal. There is only one generation in a year. As a result of the attack, the growing-point is cut off and the tree dies. The young trees are more seriously attacked by this beetle. In favourable climate the total life cycle from deposition of eggs to emergence of the beetle lasts for 4 to 9 months. Egg laying occurs practically throughout the year. The incidence of attack is highest in the dry season i.e. December to May.

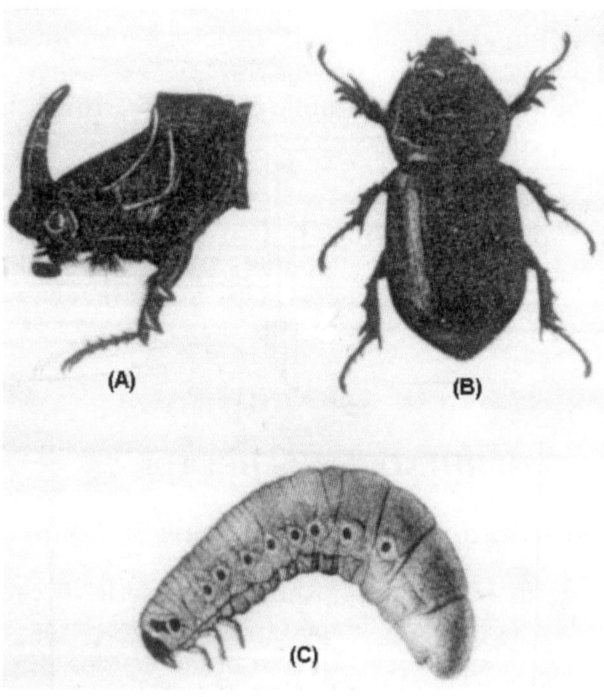

Fig. 106. Rhinoceros beetle *Oryctes rhinoceros*, **A.** Side viw of male, **B.** Dorsal view of male, **C.** Full grown larva (From Fletcher, 1914)

Control

The eggs, larvae and pupae can be destroyed by using carbaryl or lindane which kills them at the depth of about 55 to 60 cm of their breading places. The crownless trunks as well as dead trees can be cut and dried. In IPM, green muscardine fungus *Metashizium amisopliae* and *Baculobirus oryectes* is widely recommended.

2. BLACK HEADED COCONUT CATERPILLAR

***Nephantis serinopa* Meyr. (Lepidoptera: Cryptophagidae) (Fig. 107)**

Only the caterpillar is extremely harmful to coconut trees and is known as "black headed coconut caterpillar". The larval stage is a voracious feeder on coconut leaves and the damage done to the foliage is so severe that the vitality of the whole tree is lowered and the yield is very adversely affected.

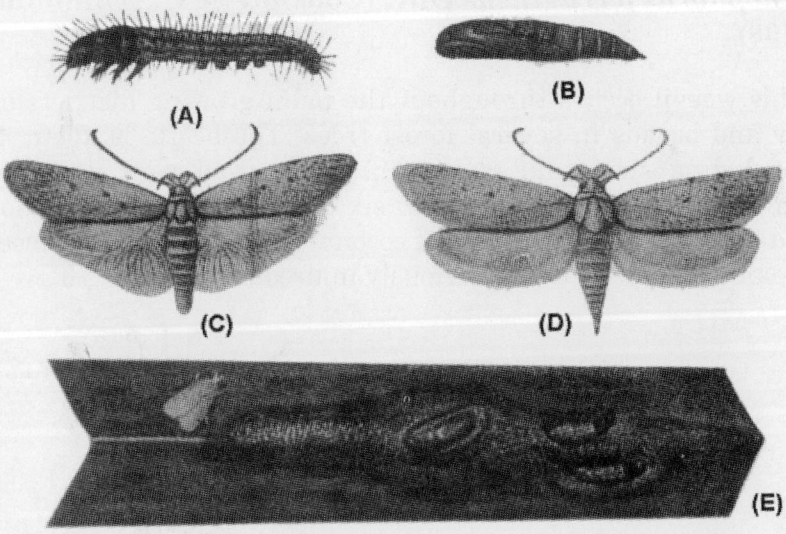

Fig. 107. Black headed coconut caterpillar, *Nephantis serinopa* **A.** Caterpillar, **B.** Pupa, **C.** Female moth, **D.** Male moth, **E.** Leaf showing galleries of the caterpillar, a cocoon and a moth (Pruthi, 1969)

Life History

Each moth lays about 100 to 125 eggs in clusters on the tips of the older leaves. The eggs are pale yellow or cream coloured but later become pink. The larva emerges in about 5 days which is extremely small about 1.5 mm. A full grown larva measures about 12 to 15 mm before pupation. The caterpillar feeds on the green matter and makes silken tunnel. About 35 to 40 days are required for full development; the larva later on transforms itself into a brownish pupa within the tunnel. In about 10 to 12 days, a moth emerges which is white in colour. The old moth is ash-grey in colour measuring about 8 to 15 mm. The damage is caused by caterpillars which construct a large number of tunnels when feeding on leaves thus causing leaves to dry. Heavy infestation causes a considerable loss of yield of coconuts.

Control

The control of this pest by releasing the hymenopterous and dipterous parasites is a simple method to check the attack of this pest on coconut trees but when parasites are used, the chemicals must be carefully managed so that the parasites must not be destroyed.

3. RED PALM WEEVIL

Rhynchophorus ferrugineus Oliv. (Coleoptera: Curculionidae) (Fig. 108)

This weevil occurs throughout the palm growing districts of the country and breads in several forest trees. The beetle is about 4 cm. long, including snout and varies in colour from very dark red to reddish brown with traces of up to six black spots on the pronotum. The red weevil is a serious pest of coconut palms especially those of 4 to 12 years old which may be seriously injured or killed.

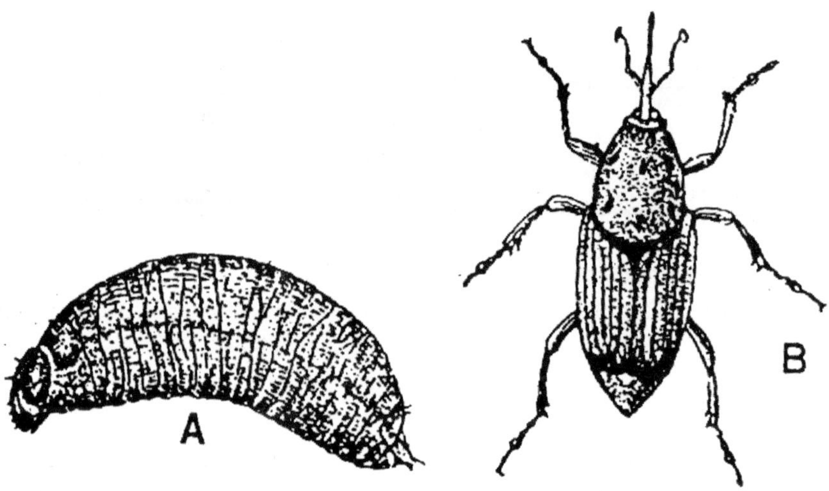

Fig. 108. Red palm weevil *Rhyncophorus ferrugineus* **A.** Larva (grub), **B.** Adult.

Life history

The female lays eggs in wounds or cracks or soft areas in the stem and bases of leaf stalks of living palms. Egg laying may begin 3 to 5 days after the beetle emerges from the cocoon and the oviposition period of a female weevil may last up to 8 weeks and 200-500 eggs may be laid. Eggs hatch in 2-5 days. The larva bores into interior of the palm stem and feeds for about a month or more on the soft tissues. When full grown, it constructs a cocoon of long stripes of fibres woven as a compact oval cell. They pupates inside the cell. The pupal stage lasts 2 to 3 weeks and immature beetle remains another week or 10 days in the cocoon before emerging. Thereafter, the weevil comes out of the tree through a hole made earlier and pluged with fibrous material. The total life cycle from egg to emergence varies between 50-112 days. Thus under favourable conditions there may be 4 to 6 generations in a year.

Control

The insects can be killed by pouring into the live holes the emulsion of dimethoate 0.5% or lindane 1.0%. Spraying on crowns every three months suppresses the pest infestation. See below for other methods.

Some other Control methods for *Rhinoceros* beetles and red-palm weevils of coconut

Beetle: A metal rod about 75 cm. long and 10 mm. diameter bent into a sharp pointed hook at one end and a ring for gripping at the other end is pushed in the beetle's tunnel and the beetle is hooked out. The cleaned out holes are filled with sand.

Fermenting oil cakes kept in water in an open vessel can be used to attract beetles to ground level.

Larva: Manure and compost heaps should be raked and spread out at least once every 3 months to destroy the larvae and to allow crows to pick out those overlooked.

Dead palms and fresh palm logs used for buildings and bridges should be destroyed if they have become breeding material and are producing beetles. Larvae boring in living palms should be killed by injecting a fumigant viz. carbon disulphide, calciumcyanide and aluminium phosphide etc. Wounds of any kind should be tarred as *Oryctes rhinoceros* makes holes which facilitates attack of *Rhyncophorus,* it should be exterminated.

24

PESTS OF SOYBEAN

It has been found that the books of general and applied entomology or agricultural entomology provide no information on the insect pests of soybean, which has been considered as a wonder crop in India. Because the soybean incorporates in itself the characteristics of pulses, oil seeds and vegetables. Thus meeting the important basic requirements of oil, protien and vegetables needed by man. Dispite numerous insects found on soybean only a small number of them are economically important. Hence, in this section an attempt has been made to describe some 8 important insect pests of soybean. The other insect pests viz. *Spodoptera, Heliothis, Diacrisia* and *Stomopteryx* etc. which are equally injurious to soybean crop, have already been described with other crop pests and therefore they have not been repeated here, the reader should consider them along with the pests of soybean.

1. GIRDLE BEETLE

Oberea brevis Swed. (Coleoptera : Lamiidae)

The beetle is distributed throughout in the various parts of India. Besides soybean, it is also a pest of cowpea (*Vigna sinensis* L.) and bitter gourd *Luffa cylinderica* etc. As a result of girdle formation, the leaf withered at the distal ends and dries up ultimately. The girdled part also dries up in course of time and breaks off.

Life history

Before egg laying, a female constructs two parallel rings or girdles on the petiole, stem or base of any one of the trifoliate leaves, usually the central one. A female lays singly about 8-72 elongate and yellowish eggs in the pith of the stem, petiole or stalk of a leaflet. Haching takes place in about 4-8 days. The larva which is yellowish white, smooth and apodous, with a swollen prothoracic segment gradually tunnels towards the main stem and hollowes it upto the base of the plant. Only one grub lives in a plant at any one time. A full grown larva measures about 19-22 mm. in length. A large number of larvae undergo diapause until the rains of the following year. Larva completes its development in about 32-62 days in non diapausing cases. In the diapausing cases, larval period lasts for about 248-308 days. In such cases, initiation of larval activity and consequently pupation depends upon the showers of rains in June-July when adults emerge in 8-11 days. Freshly developed adult is yellowish red or brown on the head, thorax and bases of elytra and lives for a few days in the tunnel itself. It comes out only when become black in colour.

2. STEMFLY

***Melanagromyza phaseoli* (Tryon) (Diptera : Agromyzidae) (Fig. 109A)**

The stemfly is a serious pest of soybean infesting about 95 percent plants in the Kharif (July-October) and up 60 percent plants in the Rabi (November –April) season. It is world wide in distribution.

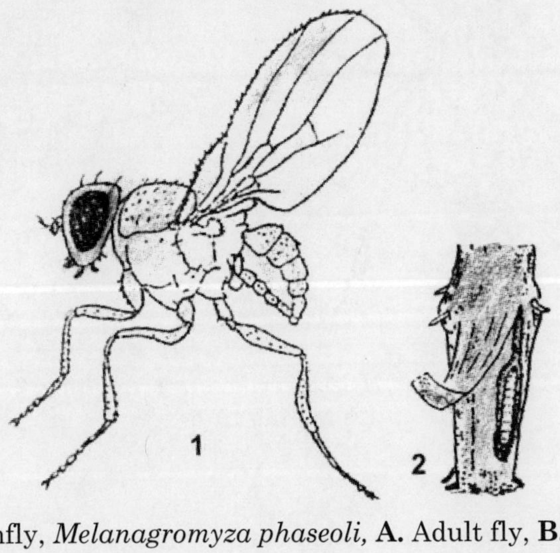

Fig. 109A. Stemfly, *Melanagromyza phaseoli,* **A.** Adult fly, **B.** Larva in stem

Life history

A female stemfly makes several punctures in the upper surface of the soybean leaf with the help of its ovipositor and lays about 14-64 eggs on the under side of the leaf below the epidermis. Inclubation period lasts for 2-3 days. The maggot enters into the stem immidiately and make zigzag gallaries upto the ground level. The maggot is yellow in colour and takes 7-12 days for full development after which it drills a small hole into the stem for future imago to escape. Only one maggot develops at a time in a stem, though 2-3 pupae may also be found at some distance from each other. The adult fly emerges in 5-9 days. There are 8-9 generations in a year.

3. SEED-CORN MAGGOT

Hylemyia cilicrura (Rond.) (Diptera : Anthomyiidae) (Fig. 109B)

The seed-corn maggot attacks soybean seeds. As a result of attack by the seed-corn maggots, soybean seeds fail to germinate, while germinating seeds fail to unfold cotyledons above the soil and dies prematurely. Besides soybean, the soybean seed corn maggot infests pea, potato, cucurbits, tobacco, cereals, cruciferous vegetation etc. It is distributed throughout the world.

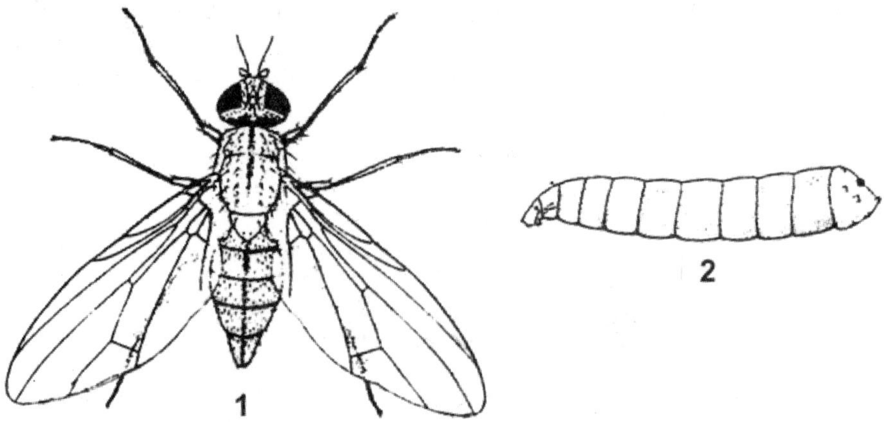

Fig. 109B. Seed corn maggot, *Hylemyia cilicrura* **1.** Adult fly, **2.** Maggot (Larva) (From Hill, 1975)

Life history

The adults of the seed-corn maggot, generally prefers germinating seeds. The female lays eggs on or near to the seeds in the soil. The eggs are elongate, white in colour which hatch in about 4-5 days in the form of small, dirty whitish maggots. The maggots feed either alone or in groups of 2-3 inside the moist swollen seeds. They are full grown in 11-14 days. Pupation takes place in the soil. The pupae are red-brown in colour and elongate. The adult flies emerge in about 7-11 days. There are several generations in a year.

4. BROWN-STRIPED SEMILOOPER

***Mocis undata* Fabr. (Lepidoptera : Noctuidae)**

The pest is distributed in several parts of the world. Extenstive damage to soybean foliage is caused by the larvae during monsoon (August- September) season, as result of which the yield is drastically reduced. The larva is slender, elongate and brownish in colour with beautiful longituidinal stripes of pink, brown and red in colour which gives almost velvety appearance to the larva.

Life history

The adult moth which is large-sized-stout and brown in colour, starts oviposition after 3-5 days of emergence and continues for about 3-8 days. The total number of eggs laid by a female moth varies from 49-204. The egg is greenish in colour and spherical in outline. Hatching takes place in 3-5 days in favourable conditions. The newly hatched larva is pale green with a pink brown head and a few setae or bristles on the body. It undergoes 6 moultings before pupation. The full grown larva measures about 41-53 mm. in length. The larval period lasts about 17-22 days after which pupation takes place within a silken cocoon, with which are attached dry leaves and soil particles. The adult emerges in about 8-17 days from the pupa. The latter is ashy dark brown in colour. The life cycle from egg to adult requires 31-35 days during monsoon (September) and 38-43 days in the post monsoon (October – December) season.

5. GREASY GREEN CATERPILLAR

***Lamprosema indicata* Fabr. (Lepidoptera : Pyralididae)**

The larva is also known as "leaf folder". The larva has the habit of folding the soybean leaves either from the tip inwards or along the mid-rib and securing the folds with the help of white resinous gum exuded by the larvae. Webbing together of several leaves is also not uncommon. The pest is distributed throughout the country but it is more common in South India.

Life history

The adult is a delicate, slender, creamy-yellow to light-brown moth with long legs. Its wings are marked with wavy irregular brown lines. A female moth lays from 54-150 eggs singly on the under side of the leaves of soybean. The egg is creamy white and flat in form. Hatching takes place in about 7-8 days. The larva is whitish green, with light brown head and a prothoracic shield of the same colour. It rolls up the leaf and grows by feeding on the chlorophyll matter only without making perforations in the leaf. As a result of chlorophyll loss, the yield is also affected. The larva is full grown in about 21-23 days, greasy green in appearance with light brown head and prothoracic plate. Pupation takes place in the folds of the leaves or in the fallen leaves on the ground surface. The pupal period lasts about 9-19 days. The total life cycle from egg to adult is completed in about 42-45 days.

6. CURIOUS FORM LOOPER LARVA

Scopula remotata Guen. (Lepidoptera : Geometridae) (Fig. 109C)

The pest is distributed throughout India. The larva of the moth appears on soybean in the second week of September to make small sieve like network of holes in the leaves. Due to their curious form and mimicry in respect of colour and appearance, they are difficult to detect. It is the larval stage which causes a lot of damage to the soybean. The adult is a medium sized, dirty-white moth, with numerous small black dots scattered on the wings.

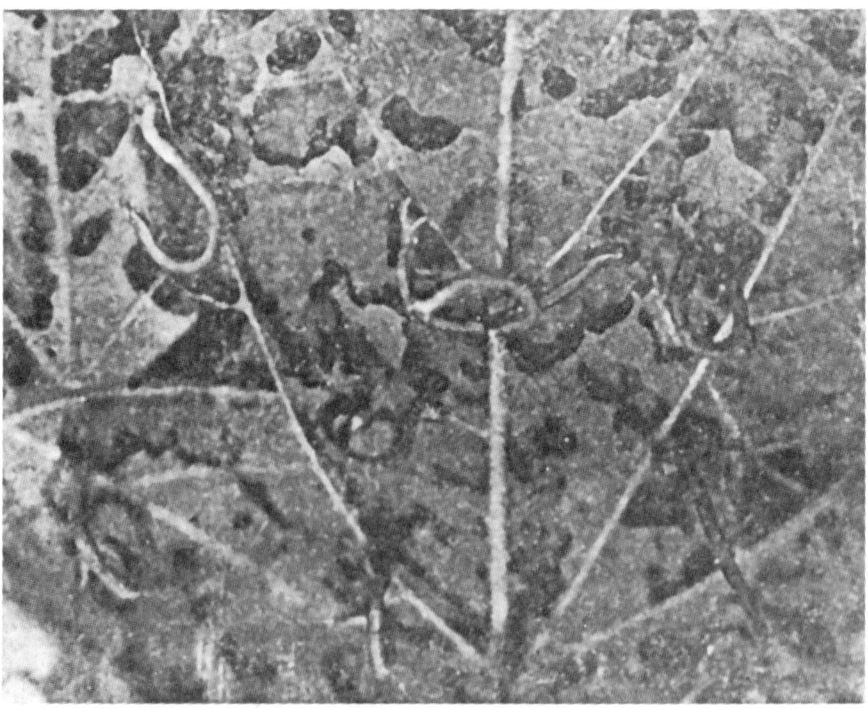

Fig. 109C. Curious forms of larvae of *Scopula remotata* on soybean leaves (From Gangrade, 1974)

Life history

The fecundity of a female moth ranges from about 28-323 eggs. The average number of eggs is about 123. Each egg is oval, yellowish green, changing to pink within a few hours, - a sign of fertility which ranges between 72-100 percent. The incubation period varies from 4-11 days and a freshly hatched larva is light brown, slender and elongate in form. The larvae of all instars are capable of twisting their body into various curious forms (See Fig. 109). They moult 5 times before changing into pupae. The full grown larva which turns green, measures about 22-24 mm. in length, which is attained in about 18-26 days in October – November and in 40-53 days in December – January. The pupa is formed in the folds of leaves. The adult moth emerges in about 8-15 days in October – November and about 15-19 days in December to January. The whole life cycle is completed in about 35-42 days in October to November and in about 59-77 days in December to January.

7. BLACK SPINY CATERPILLAR

***Syntomis fortunei* Orza (Lepidoptera : Syntomidae) (Fig. 110)**

The adults of *Syntomis fortunei* are easily recognized by their diurnal flight activity in the soybean fields, general black colour of the body and the wings. The body also bears three transverse orange colour bands on the back of the head and the abdominal segments, while the wings are transparent and covered with several white sepecks.

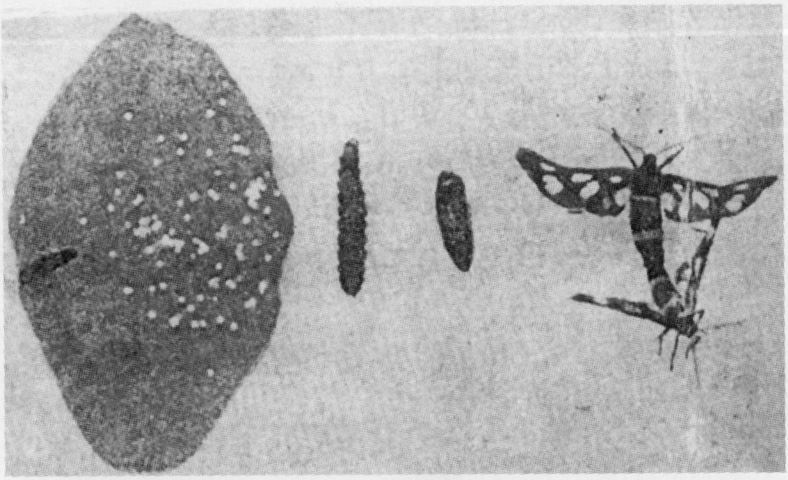

Fig. 110. Eggs, larva, pupa and adults of *Syntomis fortunei* (From Gangrade, 1974)

Life history

The fecundity of a female moth ranges from 357-548 eggs. Each egg is pale greenish in colour, spherical in shape and is laid singly or in groups. The incubation period lasts from 4-9 days. The freshly emerged larva is pink brown to black in colour with a prothoracic plate of the same colour. There are 6-7 moults before pupation. The larval period requires about 53-61 days. The full grown larva measures about 20-22 mm. in length. The pupal period requires 12-18 days after which the moth emerges.

8. BROWN COREID BUG

Riptortus linearis **Fabr. (Heteroptera : Coreidae) (Fig. 111)**

The adults apear in early September and continues through October when their activity probably synchronized with the development of the pods, and maturation of soybean. It is distributed in several parts of the world. The pest also infests rice and a wide range of plants of Graminae, Leguminacae, Solanaceae and Convolvulace.

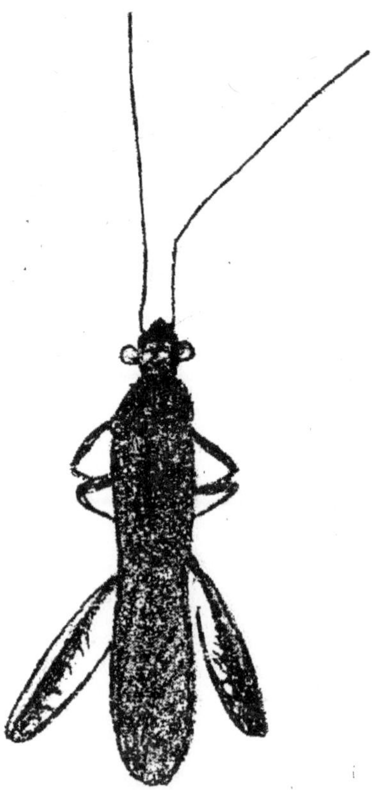

Fig. 111. Brown coreid bug *Riptortus linearis* (From Gangrade, 1974)

Life history

The adult bug is brown in colour and elongate in form with flattened hind femora bearing tooth like projections on the inner side. A female bug lays about 5-28 eggs in 11-19 days of its life. During this period the food of the bugs consist of both the leaves and green pods. The egg is brown in colour, round in shape and dorsally depressed. The incubation period lasts for about 6-10 days. The young nymph is dark brown in colour and is almost antlike in appearance and moults 5 times before becoming adult. The nymphal period is about 15-20 days. The total life cycle from egg to the adult is completed in about 21-25 days.

Control of the pests of soybean

(1) It was found that an application of lindane granules in the soil at the rate of 2.0 kg a.i. / ha. followed by two-four spray treatments in the beginning (21-30 days of germination) with monocrotophos (nuvacron) 0.04%, endosulfan (Thiodan 0.07%) or dimethoate (Rogor 0.03%) will give healthy crop. Other granules like phorate (Thimet) at the rate of 2.0 kg and 1.0 kg a.i. / ha can also be used in the place of lindane granules.

(2) Lindane dust at the rate of 0.5 kg. or 1.0 kg a.i./ha in the soil furrows before sowing followed by 2-4 sprays of ethyl parathion 46.7 EC at 0.02% proved to be the best checking the general infestation of insects on soybean including the stemfly. Thimet (phorate) 10% granules @ 10 kg/ha or disystone 5% granules @ 20 kg/ha in the soil before sowing or spraying thiodan 35 EC @ 1 litre/ha or phosphamidon @ 250 ml/ha mixed in 1000 litres of water control the pests.

(3) The infestation of soybean by the gridle beetle, *Oberea brevis* can be effectively checked by sprays of endosulfan 35 EC, methyl demeton 30 EC and dimethoate (Rogor) 30 EC in 0.05, 0.03 and 0.03 percent concentrations. Endosulfan is significantly more effective than monocrotophos 40 EC.

(4) The seed corn maggot, *Hylemyia cilicrura* can be controlled by the furrow application of phorate (Thimet) 10 gm. at the rate of 2.0 kg / ha. and dimethoate (Rogor) 5 gm., (1.0 kg a.i./ha.). The latter insectides proved no better than check (untreated). Further, carbaryl (Sevin) is very powerful insecticide for the control of many pests of soybean but it causes phytotoxicity and therefore should not be used.

25

INSECT AND OTHER PESTS OF FLORICULTURAL CROPS

The floricultural crops are mainly attacked by the insects belonging to the order Lepidoptera, Homoptera, Thysanoptera, Coleoptera and Diptera.

The most serious pests in the Lepidoptera are found in the following groups : leaf rollers, leaf tiers, plume moths, loopers, cut worms, caterpillars, sphinx moths, pierids and leaf miners. The kind of injury, generally associated with lepidopterous pests is that, the plant may be completely eaten or skeletonized by free feeding larvae, or the leaves and flowers may be first folded and then devoured, as by leaf rollers and leaf tiers.

In the order Homoptera, the pests of floricultural crops are : scale insects, mealy bugs, white flies and aphids. They may suck the sap, making the plant look unhealthy. They are also responsible for discolouration, loss of vigour and quality reduction of flowers. Numerous species of armoured scales attack garden plants the most common is the oleander scale. It has a circular shield with a central nipple.

The plants infested with mealy bugs (Coccidae) are often covered with waxy white soft sacks in which they lay their eggs. The most common species is the citrus mealy bug which is found on a wide variety of plants. The white fly (Aleurodidae) larvae suck the sap from the plants. The greenhouse white fly is the most common species in greenhouses, nurseries and infests many kinds of flower-palnts.

Aphids (Aphididae) are common pests of floricultural crops and include: green peach aphid, potato aphid, melon aphid, lily-aphid, rose aphid and tulip bulb aphid. The leaf-hoppers (Jassidae) e.g. rose leaf hoppers can be the serious pests because of their ability to transmit virus diseases such as aster yellows. The thrips (Thripidae) are of importance to florists. Defects in the flowers and leaves due to feeding of thrips is also very common.

The beetles (Coleoptera) are well known pest of floricultural crops. In most cases the adult beetle is responsible for the damage e.g. the cucumber beetles, rose chafers and blister beetles are destructive flower feeders. Flea beetles are particularly damaging to the foliage and floral parts of several crops. Adult scarabs such as rose chafer, Asiatic garden beetle and oriental beetle are general skeletonizers of the foliage and flowers, in garden flower production. Only a few species of Diptera are destructive to floricultural crops, some are leaf miners and others are seed-corn maggots.

In addition to above mentioned groups, the spring tails, termites, earwigs, lace bugs, millipeds, mites, slugs, snails, nematodes etc. are often found associated with the production of floral crops and are sometimes destructive.

1. FLOWER THRIPS

Frankliniella tritici **(Fitch) (Thysanoptera : Thripidae) (Figs. 112-114)**

Thrips are blown by wind into the green houses, nurseries and elsewhere and in this way they constitutes a complex of species that are involved in the injury caused to the floricultural crops. Their complex composition varies with location and season, the important species includes : flower thrips, tobacco thrips, onion thrips, pear thrips, composite thrips and green house thrips (Fig. 112). The damage is caused by the thrips in several ways. They may directly injure the flowers (Fig. 114) with their rasping-sucking mouth parts. They can easily puncture the surface of petals and suck the oozing juices. Their feeding produces streaks and browning of the tips of these parts. Thrips may also attack and injure bulbs and seeds.

Adults of flower thrips are small insects about 1 mm. long and 1/3 mm. wide. They vary in colour from brown to yellow. Their wings are narrow provided with fringe around the margins. Immature forms are lighter in colour, smaller in size than the adults and without wings.

Life history

The females have a preoviposition period of 1-4 days or more depending on the temeprature. The eggs are bean shaped (Fig. 113A) and are inserted into plant tissues. The eggs hatch in about 3 days and the first instar nymph requires 2 days. The mean period of second instar is 3 days, while the next instar, sometimes designated as "propupa" requires about 1 day. The "pupal" instar requires 2.5 days. The total life cycle from egg to adult, is completed in about 11-12 days. Occasionally 7-22 days are required for completion of one generation depending on temperature. In warmer parts of the country there may be 12-15 generations in a year.

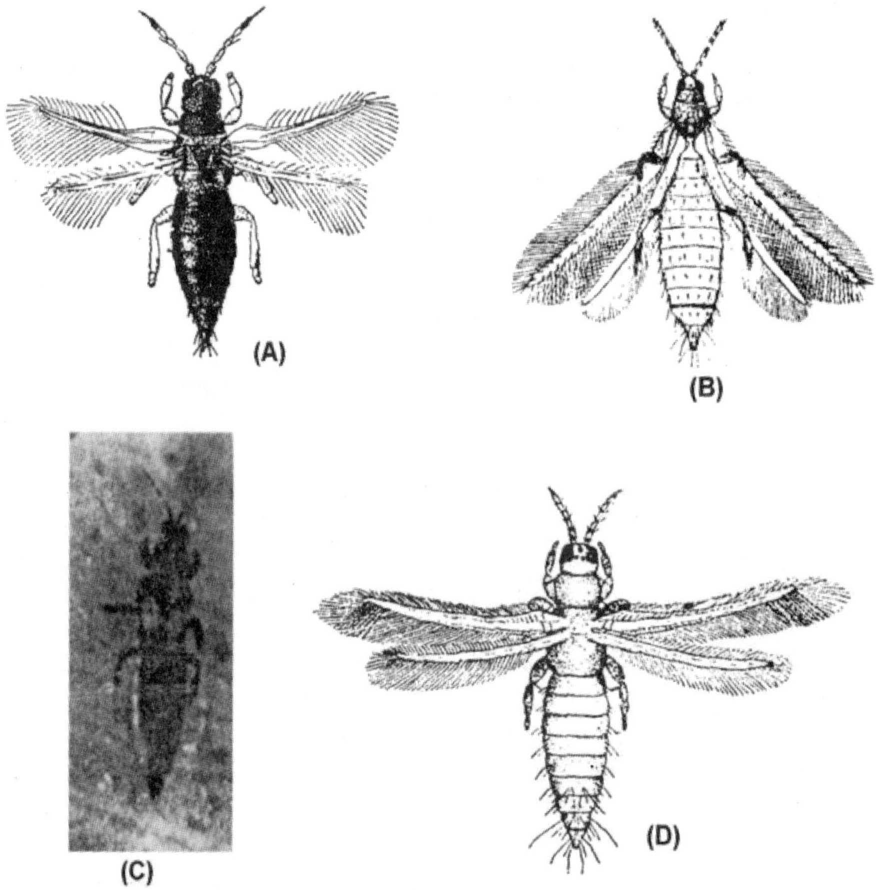

Fig. 112. Adults of four species of *Frankliniella* thrips commonly found in greenhouses, **A.** Green-house thrips; **B.** Pear thrips; **C.** Onion thrips, **D.** flowr thrips

Control

Application of insecticides such as lindane, or malathion to plants has given varying degrees of success because of differences in numbers of thrips invading the green houses or nurseries from time to time. Systemic insecticides applied to the soil in granular form or sprayed on the plants have also been used successfully for the control of thrips.

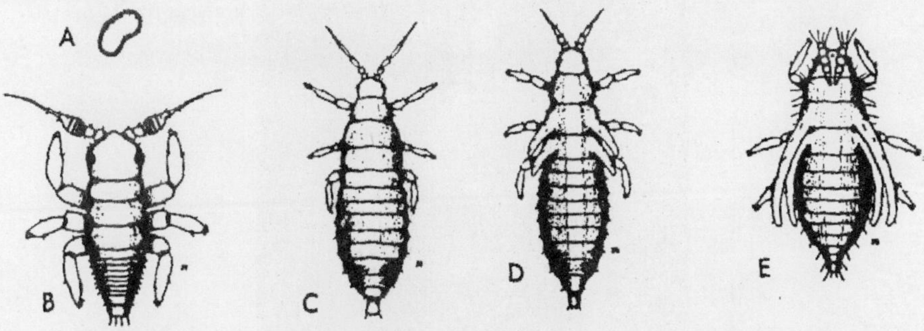

Fig. 113. Immature stages in the developmnt of thrips. **A.** egg; **B.** newly emerged nymph; **C.** full-grown nymph; **D.** first resting stage or propupa; **E.** second resting stage or pupa

Fig. 114. Thrips damage to carnations. Note the injured petal edges.

2. ROSE APHID

Macrosiphum rosaeiformis Das (Homoptera : Aphididae) (Fig. 115)

The rose aphid is distributed throughout the world. The rose aphid has been found infesting roses in winter i.e. from November to April in northern India. As a result of heavy attack the vigour of the plant is reduced and the quality of flowers deteriorates. Several other species of aphids also infest rose plants in winter. The aphid sucks the cell sap from leaves, buds and twigs. Sucking of sap from the tender buds causes disfigurement and withering of the flowers. As a result of honey dew excretion by these aphids, a black fungus may also develop, giving thereby an ugly apperance to the plant.

Fig. 115. An aggregation of rose aphids, *Macrosiphum rosaeiformis* on buds and stems of rose

Life history

The non-winged rose aphid has an elengated body, measuring about 2.5 mm., with large red eyes, black cornicles and the yellowish-geen tip of the abdomen. The species found in southern India has purple abdomen, yellow head and drak legs. The nymphal development is completed in about 11-14 days in non-winged forms and in about 14-19 days in winged forms. The rose aphid appears in the middle of November in northern India. The population increases progressively and is the highest in March, declining early in April as the season warms up. The winged forms are present throughout the period from November–April. Their number increases from December onwards which reaches at the peak in March, when about 90% of the adults become winged. It is well known that the aphid multiplies most rapidly late in the spring but cannot tolerate the summer heat. Consiquently its population declines after April.

Control

Strong insecticides should not be recommended because the tender shoots and flowers may be harmed. Spraying with 0.05% malathion / 0.03% phosphamidon / methyl demeton are very effective to control this pest. On the basis of the presence of aphids, the treatment may be repeated to check infestation.

3. TWO SPOTTED SPIDER MITE

Tetranychus telarius **Linn. (Acarina : Tetranychidae) (Figs. 116, 117)**

The two spotted spider mite is probably the most serious pest in floriculture. It attacks a wide variety of host plants. Carnations, gardenias, roses or green house and nursery crops constantly under attack. Roses are specially vulnerable because the cultural conditions provide a favourable habitat for this mite to survive and multiply (Fig. 117). Heavy infestation cuases leaves to drop and plants to lose vigour and bear a few flowers. The two spotted spider mite is less than 0.5 mm. long. Males are slightly smaller than the females. All the four life stages are covered with bristles. The body has a dark spot on each side which varies in colour from dark green to light orange depending upon the feeding activity and several other factors.

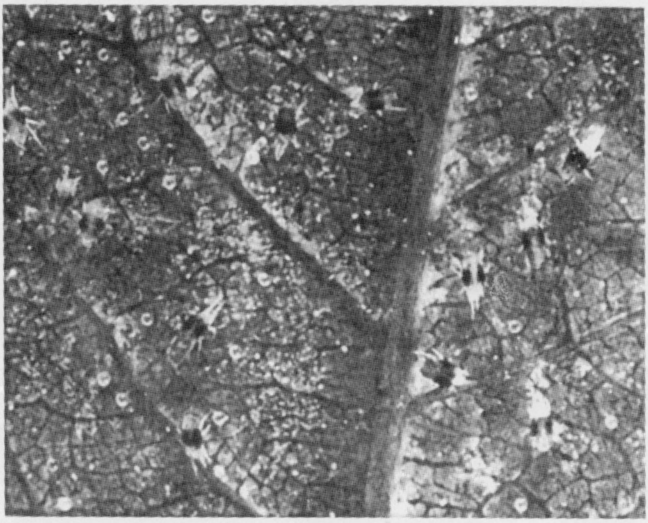

Fig. 116. Two-spotted spider mite, *Tetranychus telarius* L., adults and eggs. (From Pfadt, 1962)

Life history

The female two spotted spider mite lays 3-14 eggs per day. The average female usually lays about 70 eggs in her life time but some will lay as many as 200 eggs. The eggs are clear to pale green, spherical and about 0.1 mm. in diameter. Incubation period lasts from 2-15 days depending on temeprature. Upon hatching from the egg, the first or larval stage has only 6 legs. Depending on temperature this stage lasts from 1.5 – 4.5 days. Larvae go into a resting period and at the end of which they shed their skin and emerge as "protonymphs". This stage has 8 legs and lasts from 1.5-3.5 days. A resting period again occurs at the end of which the skin is cast and the deutonymphal or second nymphal stage appears. The mite remains in this stage from 1-16 days depending on the temeperature, with the male having a shorter deuto nymphal stage than the female. Females live for 3-4 weeks and produce eggs throughout this time. Interstingly enough, the male offspring of any female are haploid while the female offspring are diploid. The length of the life cycle varies from 5-40 days.

Fig. 117. Severly infested rose shoot completely webbed by the two-spotted spider mite *Tetranychus telarius* (From Pfadt, 1962)

Control

Certain oils and acaricides have been used to control this pest. Organophosphorus insecticide like methyl parathion has also been used to control mites.

4. SLUGS AND SNAILS

Mollusca : Gastropoda (Figs. 118-120)

The slugs and snails are characterized by having slimy soft bodies and extensible eye stalks (Fig. 118). They are mostly found in damp places. They prefer to feed on seedlings and the more succulent parts of the plants and devour flowers, leaves, (Fig. 119) stems as well as roots. In feeding, young slugs rasp away the surface tissues while older slugs make irregular holes. The difference between slugs and snails is that, the slugs are the snails, without shells. They vary in size depending on age and species. The common garden slug *Limax* spp. (Fig. 118 A-C) and *Laevicantis* spp. have been found feeding on a number of flower-plants like – balsam, marigold, dahlia, cosmos, lily etc. in the various parts of the country. Likewise, several species of the snail, *Helex* are distributed in Uttarakhand, Himachal Pradesh, Uttar Pradesh, Madhya Pradesh, Bihar, Jharkhand, Maharashtra, Orissa and other humid states of the country. The African gaint snail has been reported as a serious pest of fruits, vegetables and floricultural crops in the costal regions of Orissa, West Bengal, Assam, Tamil Nadu and Kerala.

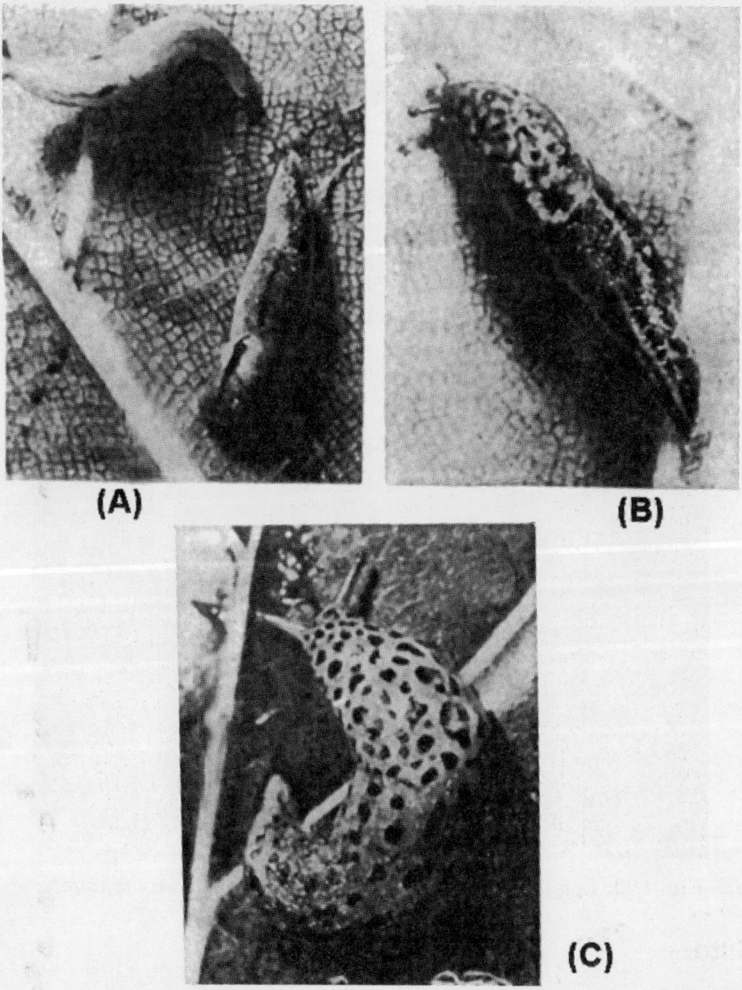

Fig. 118. Some common slugs found in green houses. **A.** *Limax marginatus,* **B.** *Limax maximus* stripped form, **C.** *Limax maximus,* spotted form: (From Pfadt, 1962)

Fig. 119. Slug injury on gerbera

Fig. 120. Egg mass of the spotted garden slug *Limax maximus*

Life history

Slugs and snails breed in the spring and the summer. All slugs lay eggs. Each species requires a different length of time for the development of its eggs and maturing of its young. Slug eggs are laid in clusters (Fig. 120) in concealed moist locations. The number of eggs laid at one sitting by one slug may approach 100 but averages 20 or 30. In spring, the eggs hatch and the slugs reach maturity and lay eggs in atumn. Most species over winter in the egg stage. Slugs reach full adult size in 3 months to a year depending on the environmental conditions. The snails also have the same breeding season. The eggs of snails hatch within 2 weeks. The shell of snails increases in size with age and are full grown in about 1-2 years depending on the availability of food and weather conditions. In the winter they stay in colonies in the rockeries, loose boards of fences, at the bottom of the hedges and rubbish heaps. Normally both the slugs and snails have only one generation in a year but when the conditions are favourable there may be a second generation too.

Control

The slugs and snails can be easily destroyed by hand picking and crushing. Regarding chemical control in flower beds or places devoid of vegetation, they can be killed by dispersing widely copper sulphate or by using metaldehyde-bait with herbaceous matter. Spraying of carbaryl 0.5% in the field at night or dusting with carbaryl 10% in day time are also effective against the slugs. Metaldehyde alone or combined with some other insecticide is the best chemical method for the control of slugs on floricultural crops. The 15% metaldehyde dust is preferable to the 20% metaldehyde liquid for slug control.

26

PESTS OF STORED GRAINS

1. RICE WEEVIL

Sitophilus oryzae Linn. (Coleoptera: Curculionidae) (Fig. 121)

It is one of the most destructive stored gain pests. These insects frequently cause complete destruction of grains and the infested grains will usually be found to be heating at the surface, and it may be damp, sometimes to such an extent that sprouting occurs. Normally the kernels of the grain are eaten. Adults feed upon whole seeds or flour but the larvae develop only in seeds or in pieces of seeds large enough for them to live within.

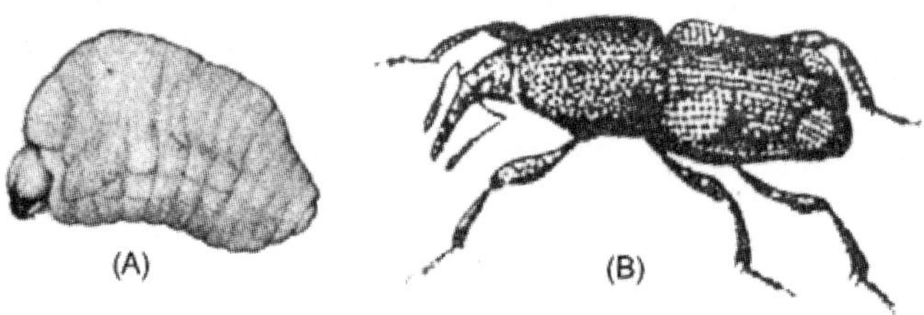

Fig. 121. Rice-weevil, *Sitophilus oryzae* **A.** Grub, **B.** Adult

Life history

The beetle in appearance is small, reddish brown in colour and is about 3.5 mm. long. The female deposits small white eggs in a hole in the grain bored by the powerful, short jaws. One female can lay as many as 250-400 eggs during its life time. The eggs usually hatch in about 3-4 days into soft white legless larvae. The larvae bore their way into the grains until full grown. In about 3 weeks to a month, the larvae pupate inside the grains and live in this stage from 3-6 days. At the end of this period the adults emerge, immediately mate and a new family is thus started. The entire life cycle of this insect, under favourable conditions, ranges from 4-7 weeks. There are 5-6 generations in a year.

2. LESSER GRAIN BORER

**Rhizopertha dominica* Fabr. (Coleoptera: Bostrychidae) (Fig. 122)

It does almost as much damage as the rice weevil. It feeds in both larval and adult stages in the interior of nearly all types of grains. It can also ingest flour. It is most common in wheat and is one of the most destructive wheat insects.

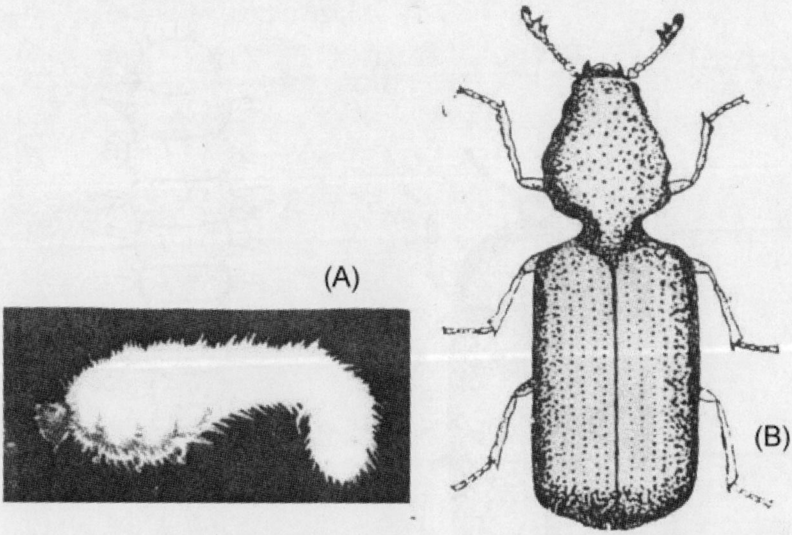

Fig. 122. Lesser grain borer *Rhizopertha dominica* **A.** Grub, **B.** Adult

Life history

The insect is dark brown-black in colour and is about 3.5 mm. long. The female lays eggs on the grains, but some times it will lay on bags, walls and among other food grains in the store. One female can lay as many as 500 eggs. The eggs hatch in 4-10 days into white larvae. The larvae are weak and are unable to attack the grain as the adult does. Consequently they search for soft corners or small holes in the grain. The larval period ranges from 40-50 days before they pupate. Duration of pupal stage varies from 6-9 days. The adult on emergence remains inside the grain for a few days. Then it cuts its way through the side of the grain and comes out. The life of adult males and females varies considerably. There are 5-6 generations in a year. The whole life history of this insect lasts about 1-2 months, depending on the temperature.

* Also written as : *Rhyzopertha dominica* Fabr. / Bostrichidae

3. KHAPRA BEETLE

Trogoderma granarium Everts (Coleoptera: Dermestidae) (Fig. 123)

This is perhaps the hardiest among the insects that infest stored grains in India. This beetle feeds on most grains and sometimes on pulses. Wheat flour is its best food and only the larvae of this insect are destructive while the adults are harmless.

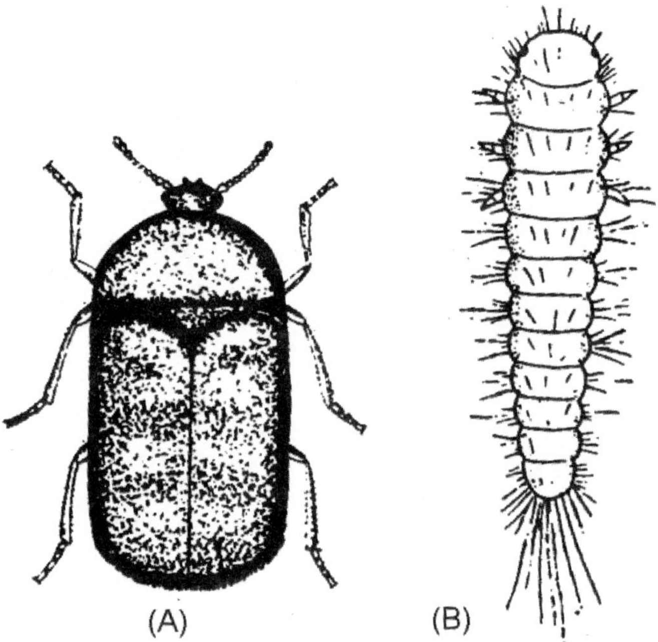

Fig. 123. Khapra beetle, *Trogoderma granarium* **A.** Adult, **B.** Larva (grub).

Life history

The adult is about 2-3 mm. long, oval in shape and has a sooty black colour. The female lays as many as 125 eggs during its life time among the grains. The eggs hatch in about a week into new brownish white larvae whose body is covered with reddish brown hairs that form a tail like structure at the end. These larvae normally feed on flour and cracked grains. The male larva takes 20-30 days and the female 24-40 days to fully fed when they are about 4 mm. long. The larvae are resistant to adverse conditions viz.- temperatures, humidity and starvation. Pupation takes place among the grains. The pupal period varies from 7-21 days. Under unfavourable conditions, the length of larval life can extend to 4 years. The adults are incapable of flying. There are 4-5 generations in a year.

4. RUST RED FLOUR BEETLE

***Tribolium castaneum* Herb. (Coleoptera: Tenebrionidae) (Fig. 124)**

It feeds mostly on broken grains or grains damaged by other pests. It is also a pest of prepared cereal products like Atta, Maida and Suji. A heavy infestation of products by these insects gives a yellow colour to the food which becomes mouldy with the excreta of larvae and adults. Both larvae and adults attack the grains.

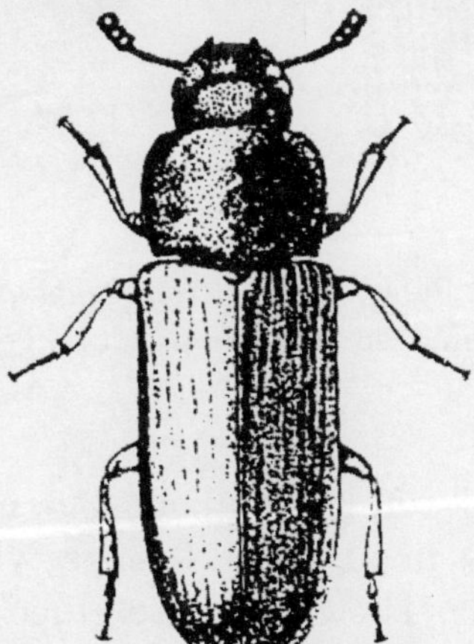

Fig. 124. Rust red flour beetle, *Tribolium castaneum*

Life history

The beetle is reddish-brown in colour and is about 3.5 mm. long. The female lays 400-500 clear white sticky eggs scattering them in the flour or food stuffs. The eggs hatch in about 4-12 days into small brownish-white larvae which become full grown in 27-90 days depending on climatic conditions and then pupate. The pupae are white in colour and the pupal period ranges from 1-2 weeks. The pupae hatch into adults which normally live for about six weeks. The pest is more active during the rainy season and the winter is passed in adult stage.

5. PULSE BEETLE

***Callosobruchus chinensis* Linn. (Coleoptera: Bruchidae) (Fig. 125)**

It is a serious pest of stored pulses. All kinds of pulses are eaten by this pest and stored pulses may be completely destroyed. It is distributed throughout the tropics and subtropics.

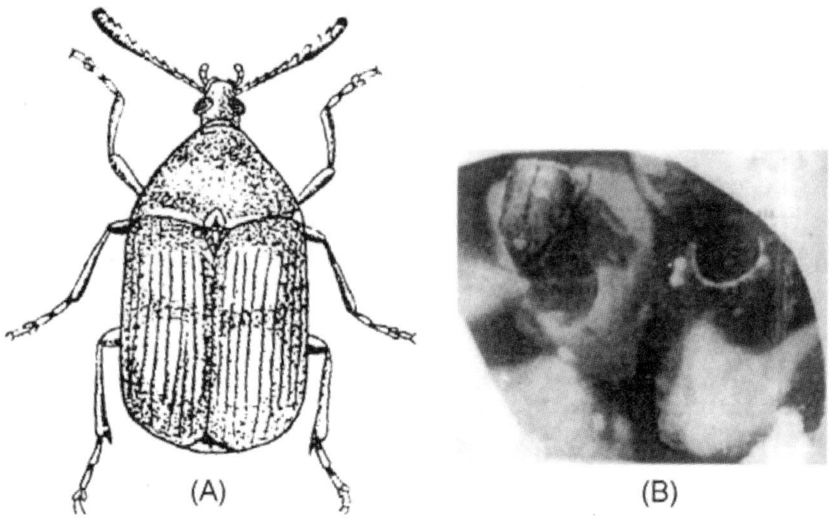

Fig. 125. A. Pulse beetle, *Callosobruchus chinensis*
B. Damaged seed and the pulse beetle

Life history

The adult is small roundish about 3.5 mm. long, brownish in colour. The female lays about 80-100 eggs which are elongate and yellowish white in colour. These eggs are glued to outside of the grain and there may be as many as 1-12 eggs on a single grain. The eggs hatch 5-18 days later into tiny-yellowish larvae which bore into the pulses. The hole made by the larva is so small that it is usually not seen. Inside the grain the larva grows slowly consuming most of the material inside. The larval duration is about 3 weeks then the larva pupates inside the grain and in about 4-7 days it comes out from the pupa as an adult. Immediately after emergence mating follows and another family is started.

6. ANGOUMOIS GRAIN MOTH

Sitotroga cerealella (Oliv.) (Lepidoptera: Gelechiidae) (Fig. 126)

This pest is commonly known as "Angoumois" grain moth "simply because it was first discribed from the French province-Angoumois. Now it is widely distributed in the tropics and some parts of the temperate region.

It is considered as an important pest of stored grain, such as wheat, maize, sorghum, barley, oats etc. and only the larvae cause damage by feeding on the grain kernels.

Life history

The adult is a small, buff coloured moth, very active in flight and is about 10- 15 mm. from wing tip to wing tip. The single female lays several hundred small, ovoid eggs, pinkish to reddish in colour, singly or in clusters of as many as 20 eggs on the grain. The eggs hatch in about a week and the tiny larvae about 8 mm. long immediately bore into the grain. The larval period ranges from 2-3 weeks, after which pupation starts inside the grain in a silken cocoon. After a pupal period of about a week within the grain they emerge as adult moth which do not feed on grains but live for a few days only. Breeding continues throughout the year except during the cold weather. The whole life cycle takes about 5 weeks at 30°C. There are several generations in a year.

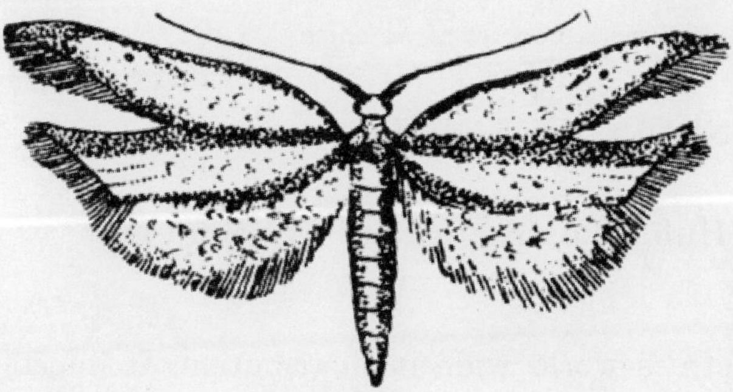

(A)

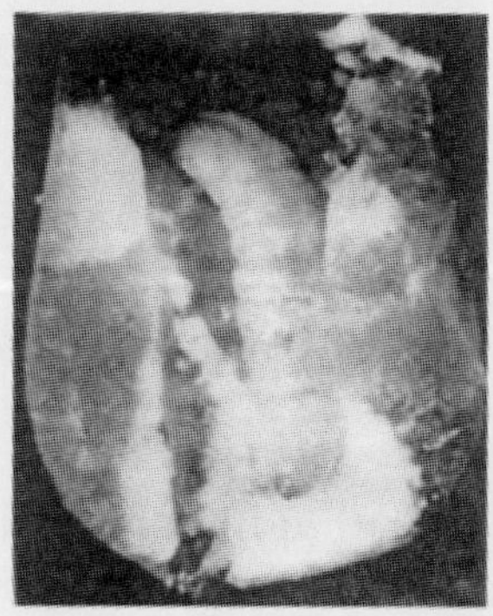

(B)

Fig. 126. A. Angoumois grain moth, *Sitotroga cerealella* **B.** Larva of the Angomois grain moth feeding in a kernel of corn.

7. RICE MOTH

Corcyra cephalonica Staint. (Lepidoptera: Galleriidae) (Fig. 127)

It is a serious pest of stored paddy and rice. Besides feeding on rice, it also infests gram, maize, sorghum, groundnut etc. It is slightly larger than the grain moth. Only the larvae damage the rice-grains by feeding under silken webs. When infestation is severe, lumps of grains may be convered into a webbed mass.

Life history

The female lays as many as about 200 eggs loosely on the bags, walls or grains. The eggs hatch in 4-5 days into tiny creamy white larvae which become fully grown in about a month. At the end of this period the larvae spins a tuff silken cocoon and pupates inside this. The pupal period varies from 6-10 days or more depending on temperature, after which the adult emerges and breeds. There are about 5-6 generations in a year.

Fig. 127. Rice moth, *Corcyra cephalonica*

8. INDIAN MEAL MOTH

***Plodia interpunctella* (Hub.) (Lepidoptera : Phycitidae) (Fig. 128)**

The Indian meal moth is world wide in distribution. It infests grains, meals, breakfast-foods, soybean, dried fruits, nuts, seeds etc. The damage is caused by the larvae. In the early stages they appear whitish pink or green in colour.

Life history

The female moth lays about 30-350 eggs oval in shape and whitish in colour. They are laid singly or in clusters on or near to the appropriate food materials. The incubation period is 2 days to 2 weeks depending on the climatic conditions. The mature larva is about 8-13 mm. in length. The larvae feed upon grains or other foods and become full grown in about 30-35 days. Pupation takes place within a thin silken cocoon. The pupal period lasts 4-35 days depending on the weather conditions. The adult moth is about 13-20 mm. in wing expanse, with a coppery lustre on the outer two-third and is whitish grey on the inner portions as well as on the end of the body. In summer the life cycle is completed in about 5-6 weeks. There are about 4-6 generations in a year.

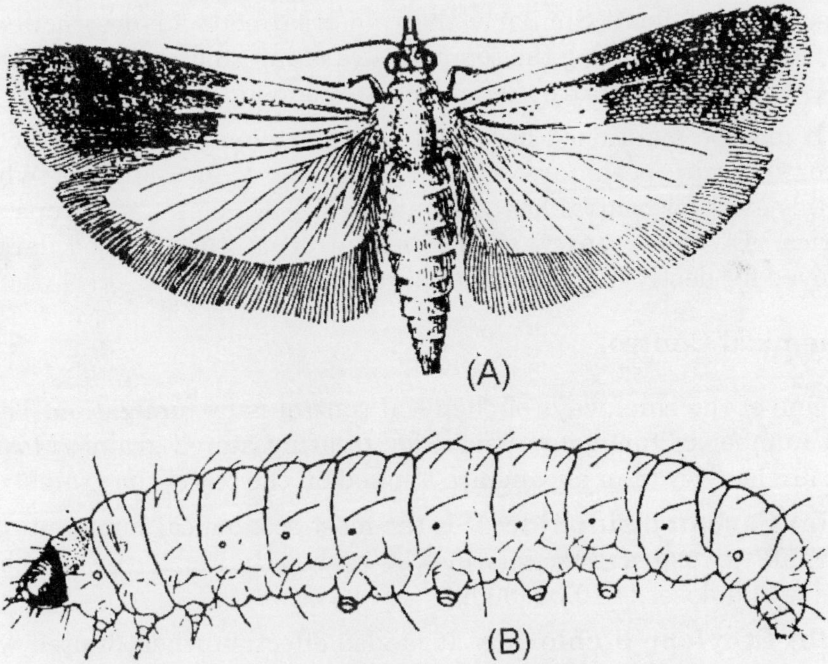

Fig. 128. A. Indian meal moth *Plodia interpunctella* and **B.** Larva

CONTROL OF STORED GRAIN PESTS

Control of stored grain pests has been a subject of extensive study for a very long time. The methods adopted in recent times to control these insects have been very simple and very effective. Generally all these insects may be destroyed by heat and cold treatment as well as by treatment with fumigants, powerful contact insecticides, in smoke, fog or spray forms or by desiccating or by use of abrasive dusts.

1. Destruction by heat and cold treatment

Heat treatments generally give a good result in mills, stores, store rooms, houses etc. with the result that practically all the insects are killed. The expense involved in such a method of control is lower than that of any other method. Heat treatment of any place infested by stored grain pests should normally be carried out during the period when the outside temperature is high. This method essentially consists of heating up the entire room by a system of tubes through which hot water is continuously circulated. The hot water from these tubes should be able to maintain a temperature ranging 49°-66°C for several hours. Such a range of temperature is fatal to all insects exposed to this heat. Similarly the cold treatment for destruction of insects involves lowering the temperature of infested room to as low as 0°C. This low temperature is equally fatal to insect life.

It may be mentioned here that the above methods of destruction of stored grain pests are normally practiced in countries where facilities exist for central heating and cooling systems e.g. Europe and America. In our country chemical control is the method largely employed for destroying the insect pests of stored grains.

2. Chemical Control

One of the chief ways of chemical control is by fumigation. There are a number of fumigants useful for treating stored grains. Most of them are heavier than air and are applied on the top of the grains.

(a) Carbon disulphide: It is the most economical fumigant. It is preferable to use a 20% mixture in carbon tetrachloride which is effective at a doses of 10 kg./300m.

(b) Ethylene dichloride: It is also effective when diluted with 25% carbon tetrachloride. It is used as 20 kg./300m.

(c) Methyl Bromide: It is another effective fumigant. It is effective at a level of 500g./300m. It is outstanding for bagged materials such as flour on account of its great penetrating power.

(d) Chloropicrin: It is another important fumigant which is usually applied at 2.5 kg./300 m.

(e) Aluminium phosphide: It is widely used as fumigant for stored grain protection, being marketed in the form of tablets.

There are some less volatile fumigants such as ethylene dibromide and acrylonitrile which give excellent control of insects in granaries.

Stored grain pests are also controlled by vacuomax fumigants. Acrylonitrile is best used for this purpose at a dose of ½ kg/1000 ft. This fumigant is almost as toxic as hydrogen cyanide and is much more penetrating. Some other fumigants viz. ethyl dibromide are general fumigants much less toxic than methyl bromide. It is readily absorbed as a grain fumigant. The breads made from the grains have no bad odour or taste, due to its low vapour pressure and high toxicity, the fumigation of entire godown kills insects both in the grains as well as godown.

Often residual wall treatments with pyrethrum sprays have been found to have surprisingly lasting effect especially when they are deposited in cool, dark warehouses, the toxic effect remaining as long as 100 days.

A novel method of proofing grains against insects is the addition of abrasive powders such as mineral oxides or silicates which scar the cuticle and cause desiccation of the insects e.g. certain abrasive materials are magnacite, silica, hydrated lime, carborandum etc.

Many insecticides have been used as impregnants for food packages to prevent infestation by stored product insects. They include rotenone, fixed nicotine, etc. Sometimes these insecticides contaminate the food stuff itself and render them useless for consumption. Consequently, in recent times, preference has been given to certain inorganic substances such as NH_4Cl which functions as gustatory repellent.

Another method of protecting grain from insects attack is by the use of grain treatments which are applied either as sprays or as powders to uninfested grains. Some of the modern grain protectants are pyrethrin synergists piperonyl butoxide, malathion (1%). Grain protectants have certain advantages over fumigants in that: (a) the grain can be treated while it is uninfested, (b) protectants are effective when the grains are stored in containers too loose to be fumigated, (c) protectants are less dangerous than fumigants and (d) protectants do not effect germination adversely. Various parts of the neem plant have been found very effective in controlling almost all stored grain pests (See : Insecticides of plant origin).

27

PESTS OF SPICES, NARCOTICS AND DRUGS

Although a number of plants contain and produce highly effective insecticides yet they have their own serious insect pest problems. One of the best example is afforded by tobacco the source of the insecticide nicotine, is subject to devastation by the tobacco caterpillar, *Prodenia litura,* the cut worm *Laphygma exigua,* the gram caterpillar *Heliothis armigera* and the cigarette beetle, *Lasioderma serricorne.* The turmeric and ginger both are seriously infested by *Dichocrocis punctiferalis, Udaspes folus, Lasioderma serricorne* under various field or storage conditions. Likewise, the *Longitarsus nigripennis* is a serious pest of peppervine and green chillies. The chilli fruits are also attacked by capsid bug, *Disphinctus polistus.* The plants of the genus *Chrysanthemum,* the source of pyrethrum are attacked by hairy caterpillars, aphids and termites. Likewise, Cinchona the source of "quinine" is infested by a number of beetles and their grubs. Moreover, it has been observed by the author (Awasthi, unpublished) that the bark-stem of Neem plant, a source of several insecticides is badly infested by termites, *Odontotermes* spp. The mint plant the source of Menthol a chemical of great pharmaceutical importance is infested by *Laphygma exigua, Autographa (Plusia) nigrisigna* and *Diacrisia oblequa.* The Khursasani Ajvayan is infested by *Heliothis armigera, Aphis gossypii* and *Myzus persicae.* The plant Serpgandha *(Rauwolfea serpentina)* is attacked by the larva of the sphingid moth, *Dielephia nerii.* Likewise, the coriander i.e. Dhania is infested by *Spodoptera exigua, Bemisia tabaci* and *Hyadaphis coriandri.*

In the following pages account of some of the pests of spices, narcotics etc. is given and many others have already been described under the pests of specific plant species.

1. CIGARETTE BEETLE
Lasioderma serricorne Fabr.
(Coleoptera : Anobiidae) (Figs. 129, 130)

It is commonly known as the cigarette beetle and is a polyphagous pest which infests a vast variety of commodities for example turmeric, ginger, opium, tobacco, saffron, redpepper, cumin-seed, nux vomica seed, belladonna etc. This pest is distributed all over the world. The adult beetle is small and oval, light to dark brown in colour and measuring about 2.0-3.7 mm in length. But its size depends on the type of food, temperature and humidity encountered during development. It is active during afternoon and night. It is phototactic. The longevity of the adult beetle is 4 or 5 weeks or more depending on certain climatic conditions.

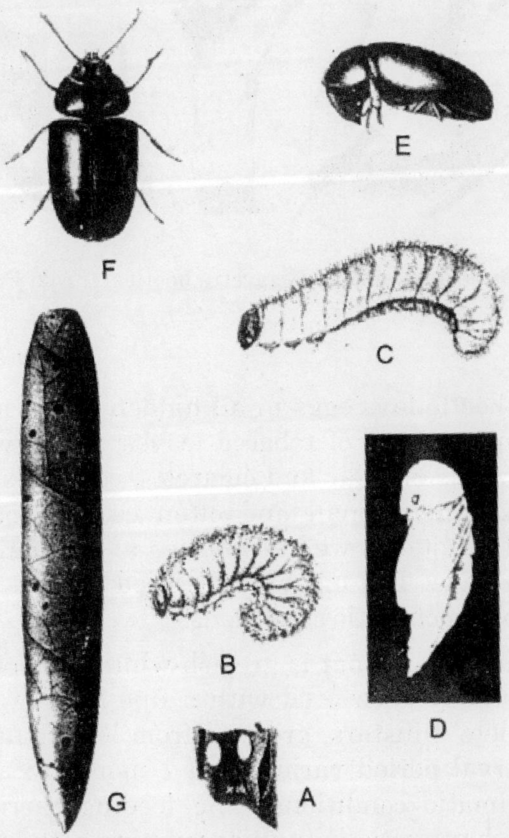

Fig. 129. Cigarette beetle, *Lasoderma serricorne*
A. Two eggs on a piece of tobacco leaf, **B.** Larva as it is usually covered with particles of leaf, **C.** Larva divested of covering, **D.** Pupa, **E.** Imago, lateral view, **F.** Imago in resting attitude, **G.** Bored cheroot. (From Lefroy, 1909)

Fig. 130. Turmeric damaged by Cigarette beetles, (From Pradhan, 1969)

Life History

The female beetle lays eggs in all hidden safety niches like - the space in between the folds of tobacco leaves, in between grains like coriander, in the folds of cigar and cigaretts, corners and crevices etc. The eggs are extremely small and often escape notice. These are pearly white and opaque as well as pointed at one end, 0.4 to 0.5 mm long and 0.2 mm wide. The incubation period may be 4 to more than 19 days depending on climatic conditions.

The larva on hatching out is greyish-white in colour, about 1 mm in length and its body is covered with a fine light-brown hairy coat. The larvae complete 4-instars, growing from less than 1.0 mm to 4.5 mm long. The larval period varies from 1 month to about 6 months depending on climatic conditions. The feeding carried out by the larvae results in characteristic tunnels running both longitudinally as well as across the tobacco leaves, cigars etc. When full grown, the larva makes a flimsy cell (cocoon) in which it pupates. The fresh pupa is whitish in colour but it turns brownish later on. The pupal period varies from 5-21 days or even more, depending on climatic conditions.

Adult beetles remain in the cell to harden and become sexually mature. Emergence from pupal cell is mainly during the night. Egg to adult development depends on food, temperature and humidity.

Control

It has been found that almost all stages of the pest can be killed by about 15 minutes exposure to 60°C. Suitable fumigation can also be used keeping in view that the fumigant doesnot react adversely on the commodity which is so varied in nature.

Pheromone traps have also been used for the control of *Lasioderma,* the essential element of the trap is a lure permitting controlled release of sex pheromones. The lure is mounted on a sticky board that catches the beetles. For *Lasioderma* three lures are commonly used: the New Serrico, the Lasiotrap and the Trece lures (Ryan, 1995).

2. TOBACCO MOTH
Ephestia elutella (Hub.)
(LEPIDOPTERA : PYRALIDIDAE) (Figs. 131, 132, 133, 134)

The tobacco moth *(Ephestia elutella)* larvae feed on a wide variety of stored products including tobacco. Adults are small, grey or brown-grey moth mesuring just under 1.0 cm from head to the tips of the folded wings. Wings have a spread of about 1.6 cm. The moths are active at dusk. In the warehouse, adults live upto 3 weeks with males rather shorte-lived than females. Copulations takes place 1 or 2 days after emergence.

Fig. 131. Tobacco moth, *Ephestia elutella* adult, dead and displayed to show the wing pattern. At rest the moth folds the wings over the abdomen. (From Ryan, 1995)

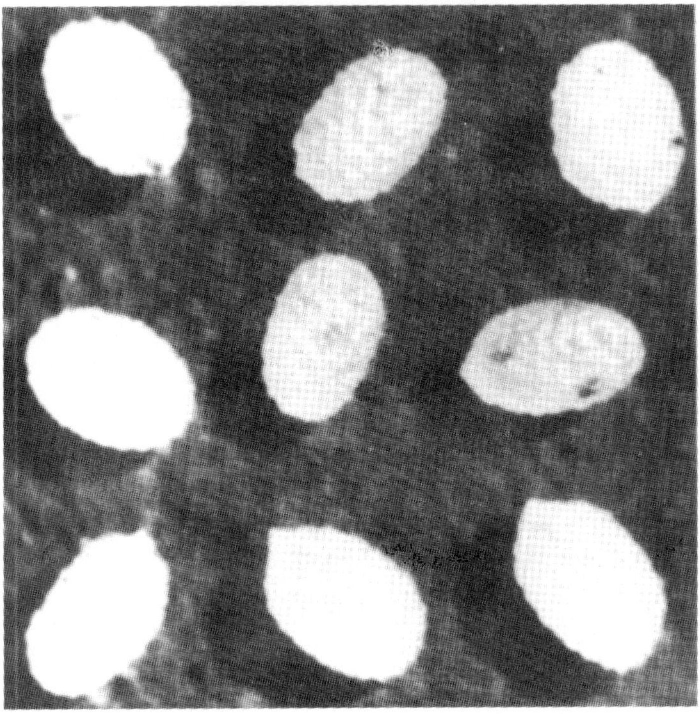

Fig. 132. Tobacco moth, *E. elutella* eggs. (From Ryan, 1995)

Life history

The females lay more than 100 eggs mostly at dusk or during night in the first 4 or 5 days of life. The eggs are laid singly or in groups on or near to the food source. They are sandy white when laid, turning darker as the embryo develops. Each egg is ellipsoid and measures about 0.5 mm long and 0.35 mm wide. The egg shell is tough, finely ridged and only loosely attached to the substrate. The time taken for egg to hatch varies with temperature but at 25°C. the larvae emerge within 6-7 days. The larvae vary in colour between light yellow, brown, pink or off-white. Larvae generally develop through 6 instars and are about 13 mm long when fully grown. After hatching larvae burrow into the tobacco, begin feeding and remain unseen for 2-3 months. Only the larvae feed on tobacco attacking primarily flue-cured and Oriantal tobacco. In the severe infestation, entire leave except for the mid-rib and larger veins may be devoured. Full grown larvae enter a wandering phase, spinning silken threads behind them as they wander and emerge from tobacco. In a warehouse the larvae often move via the walls to the ceiling or a crevice, where they spin a silk cocoon and may diapause for 4-9 months. The six instar larvae eventually pupate. The pupal colour changes from light to dark brown with the pupae becoming darker as they get older. The duration of pupal stage is governed by temperature and varies from 45 days at 15ºC to 10 days at 30ºC. Pupae are always found on the external surface since the emerging adult moth is fragile and unlike *Lasioderma serricorne*, cannot push its way out of the confinement.

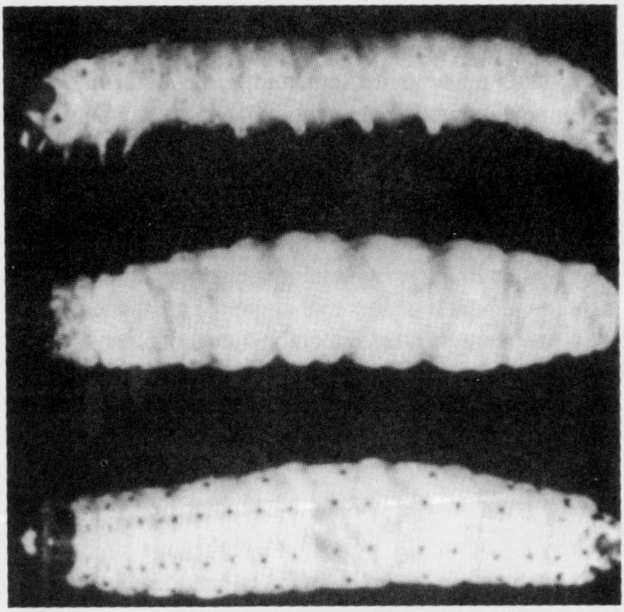

Fig. 133. Tobacco moth, *E. elutella* larvae. (From Ryan, 1995)

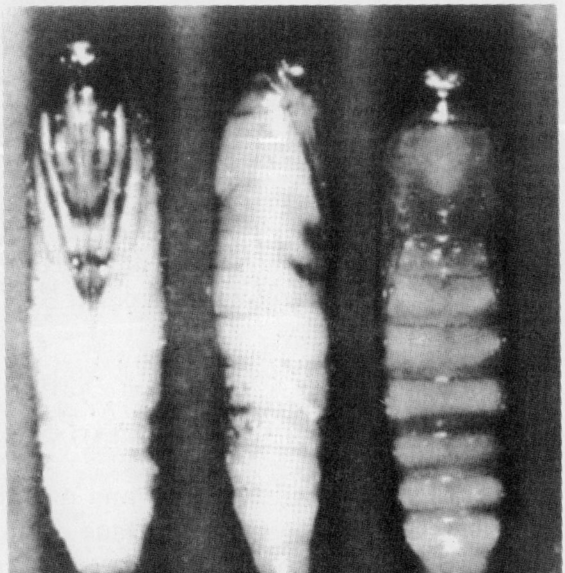

Fig. 134. Tobacco moth, *E. elutella* Pupae. (From Ryan, 1995)

Control

The discovery and synthesis of the female produced sex pehromone ZETA led to production and use of numerous bait traps for *Ephestia* spp. and the Indian mealworm *Plodia interpunctella.*

The synthetic compound mimics the effect of the juvenile hormone, hence it may also be termed as juvenoid. Insect growth regulator IGR does not eliminate the insect in the manner of conventional insecticides. The mode of action is to prevent the larval form of the insect from becoming an adult.

Bracon hebetor, Tricogramma sp. and several parasitic wasps act as natural enemies and control the pest population of *Ephestia* spp. (See : Rayan, 1995).

3. TURMERIC SKIPPER
Udaspes folus (Cr.) (LEPIDOPTERA : HESPERIIDAE)

The turmeric plant is damaged by the catterpillars of the skipper butterfly which fold the leaves and feed from within. It is widely distributed in India and several other countries. The adult skipper butterfly is purplish black with white spots on the wings. Antennae are apically prolonged beyond the club into a hook or small recurved point. It is also a pest of ginger.

Life history

The female butterfly lays eggs on the leaves of the turmeric plant. Caterpillars are green in colour with a stout body and a black head. Pupation takes place within the leaf folds in a fluffy cocoon and after completion of development adult emerges.

Control

The pest seldom causes economic damage. Hence, no specific control measure is generally required for this pest.

4. LEAF BEETLE
Lema praeusta (Fabr.)
(COLEOPTERA : CHRYSOMELIDAE)

This beetle is commonly found in Orissa and neighbouring areas. The adults and grubs both feed on the leaf tissue of turmeric, brinjal and cucurbits. The adults are small 5.6 mm long. Elytra are dark brown in colour with apical one third portion blue in colour.

Life history

The female beetle lays eggs singly. The eggs are thrust in leaf tissue. Incubation period lasts for about 8-10 days. The grubs are small 6-8 mm long, yellowish-white in colour with a black head. The anal end projects inwards and the excreta falling on the dorsal surface serves as a protective covering for the grub. The grub period lasts for about 10-12 days. Pupation takes place in the soil or between the leaves under the covering of faecal matter. Emergence of adult takes place in about 15-25 days depending on climatic conditions.

Control

The simplest method of control is hand picking of the grubs. Regarding chemical control, spraying with dichlorvos 300 ml/ha or endosulfan 1000 ml/ha is very effective.

5. BANANA LACEWING BUG
Stephanitis typicus Dist.
(HEMIPTERA : TINGIDAE)

This lacewing bug is a small insect with black body, measuring about 4 mm in length. The wings of the bug are transparent, shiny and reticulate in nature. The thorax bears lateral expansions extending to the side of the head. Nymphs and adults both occur in colonies on the lower surface of the leaves of turmeric, banana, cardamom etc. Infestation results into dwarfing and distortion with yellow patches and necrotic spots. Some times the pest causes severe damage.

Life history

Eggs are laid in the parenchymatous tissue of the leaves. Nymphs upon hatching suck plant sap and inject toxic saliva in the tissue. In warmer climatic conditions there may be several generations in a year. One life cycle is completed in several weeks.

Control

Spraying of fenitrothion @ 600 ml/ha or phosphamidon @ 300 ml/ha is extremely effective to control this pest. Spray should be particularly directed to the lower surface of leaves, since the nymphs as well as adults occur there.

6. GREEN PEACH APHID
Myzus persicae (Sulz)
(HOMOPTERA : APHIDIDAE) (Fig. 135)

This aphid is cosmopolitan in distribution and is polyphagous. It attacks the plants belonging to the families Cruciferae, Solanaceae and fruit trees like citrus and peach. The plants of great medicinal value i.e Khurasani Ajvayan as well as capsicum are infested by the *Heliothis armigera, Aphis gossypii*, and *Myzus persicae.*

Damage is caused by both the nymphs as well as adults which are oval and soft bodied, and are found on the underside of the tender leaves and shoots of Ajvayan and chillies. They suck the cell sap from the tissues. With the result leaves curl and dryup. Due to secretion of honey dew, sooty mould develops on the leaves which also inhibit growth.

The adult aphid is small in size measuring about 1.5-2.5 mm. The body colour is green with dark green thorax. The honey dew tubes, (cornicles) are fairly long. Both nymphs and adults suck the sap of Ajvayan leaves. It also transmits virus desease of plants.

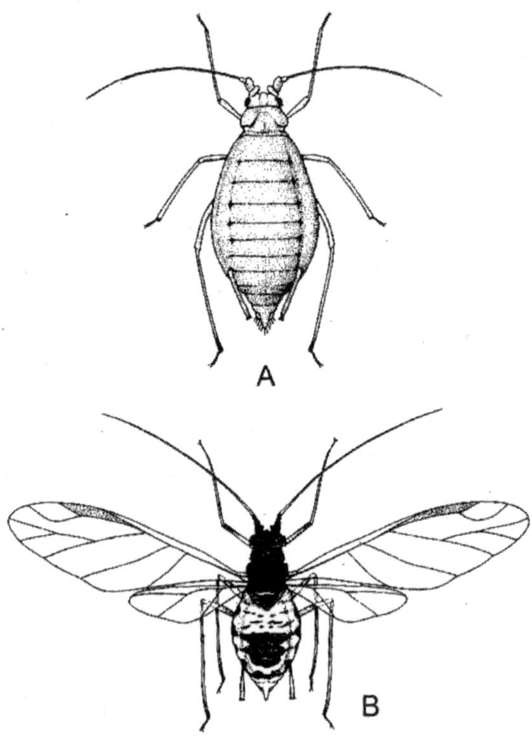

Fig. 135. Green peach aphid - *Myzus persicae*
A - Apterous adult female, **B** - Alate adult female (From Hill, 1975)

Life history

In the tropical climatic condition no males occur i.e. sexual forms are never produced. The females breed by parthenogenesis and viviparously and nymphs mature in 1-2 weeks. A female aphid may give birth to young ones for about 2-weeks, in her life, and produces about 12-14 nymphs per day and these nymphs mature and start reproducing in a week's time. In the cloudy weather, rapid multiplication takes place.

Control

Spraying the crop with dimethoate @ 600 ml/ha or phosphamidon @ 300 ml/ha or methyl demeton @ 800-1000 ml/ha is effective to control this pest.

7. COMMON CUTWORM
Euxoa segetum Schiff.
(LEPIDOPTERA : NOCTUIDAE) (Fig. 136)

This is a polyphagous, cosmopolitan cutworm of sporadic importance on a wide range of crops, occasionally the damage by this pest may be devasting. Seedlings are cut through above the ground level, one caterpillar my cut through a number of seedlings in one night. Root crops may be deeply gnawed, often at level well below the soil surface. Besides other plants, it has also been reported as a pest of mint plants (Pradhan, 1969).

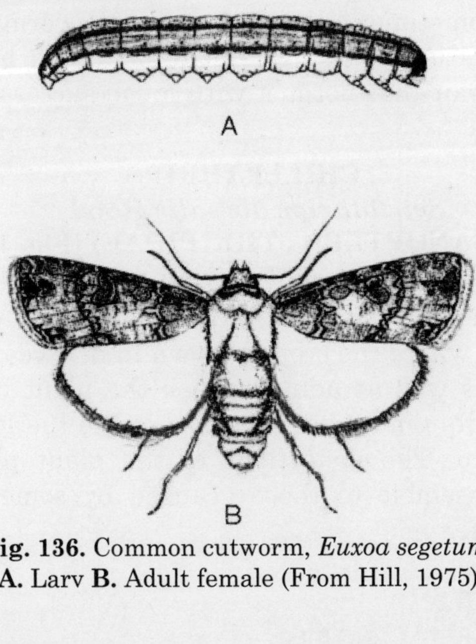

Fig. 136. Common cutworm, *Euxoa segetum*.
A. Larv **B.** Adult female (From Hill, 1975)

Life history

The eggs are laid on the stems of crop plants or on the soil and they hatch in 10-14 days. Each female may lay up to 1000 eggs.

The early instars tend to remain on the foliage of the host plant for 1-2 weeks, then as the caterpillars develop they gradually move down into the soil and assume the cutworm habits. They usually remain in the soil during the day and come to the surface to feed at night. Sometimes the older caterpillars may tend to remain in the soil all the time feeding on the root crops. The grey-coloured mature larva is 30-40 mm long and with faint dark lines along the side of the body.

The pupa, is smooth, shiny-brown with two spines at the rear, 20-22 mm long. Pupation takes place in the soil. The adult moth is 30-40 mm across the wings, the forewings are grey-brown or black with dark brown kidney shaped markings; the hindwings are almost white in the male but darker in the female. There are several generations in a year depending on the climatic conditions.

Control

Chemical control by using fenitrothion, trichlorphon and carbaryl is quite effective provided they are put on at high volume when the caterpillars are young. The older caterpillars are notoriously less susceptible to pesticides, and when they are in the soil it is difficult to achieve contact with insecticide. Fenitrothion has generally given the best control of many cutworm species.

Baits are sometimes used against cutworms either as a proprietary formulation or using fenitrothion with bran bait but in general the use of baits has not met with much success.

8. CHILLI-THRIPS
Scirtothrips dorsalis **Hood**
(THYSANOPTERA : THRIPIDAE) (Fig. 137)

This is the chief pest of chillies in all parts of India but is particularly severe where the crop is grown in dry weather. The thrips both, the youngs as well as adults attack the plant in all its stages, suck the sap from its tender portions and cause the leaves, buds and fruits to shrivel up. The distortions of the plant produced by the feeding of pests resemble to those caused by some leaf-curl virus desease or deseases.

Life history

Reproduction is both, sexual and parthenogenetic. In case of sexual reproduction oviposition period lasts for about a month. About 100 eggs, which are very minute are laid just under the leaf epidermis @ 2-4 eggs/day. The young ones on hatching crawl on to the tender shoots for feeding. After the second moult, these nymphs enter into a short semi-quiscent pre pupal stage and then pupate in sheltered places such as the leaf axiles, leaf curles and the base of flowers and fruits. After 3-5 days the adults emerge, which have heavily fringed wings that are uniformly grey in colour. The whole life cycle is completed within 10-15 days. There are several overlapping generations in a year. This thrips is also found on several other plnts like bean, castor, groundnut, mango etc. The intensity of infestation on chillies at the same time of the year varies in different parts of the country and is mainly influenced by the prevailing weather conditions. Rainfall keeps the population down. The pest thrives best and multiplies rapidly in warm but not very dry weather.

Fig. 137. Chilli thrips, *Scirtothrips dorsalis*
(From Srivastava and Butani, 2009)

Control

The pest can be effectively controlled by dusting with 1% rotenone and 170 kg/ha is enough. Malathion 1% or lindane sprays 0.01% are also effective. In the earlier stages spraying with nicotine sulphate (1 : 1000) should be adopted. This thrips can also be controlled by spray with 0.03% dimethoate or monocrotophos.

9. COTTON APHID
Aphis gossypii Glov.
(HOMOPTERA : APHIDIDAE) (Fig. 138)

This aphid is worldwide in distribution. It is polyphagous in nature feeding on a variety of plants belonging to families Malvaceae, Cucurbitaceae, Leguminosae etc. In addition to cotton, it is also an important pest of capsicum (chillies). The aphid is deep-brown to greenish-black in colour. The antennae are usually half as long as the body and the eyes are red. The cornicles (siphonculi) are black. They live in colonies and suck the sap from the tender part of the shoots and underside of the leaves. Drainage of cell sap results in leaves curling up, like a cup. Tender portions become blighted. Honey dew secreted by the aphid encourages sooty mould on the upper surface of leaves.

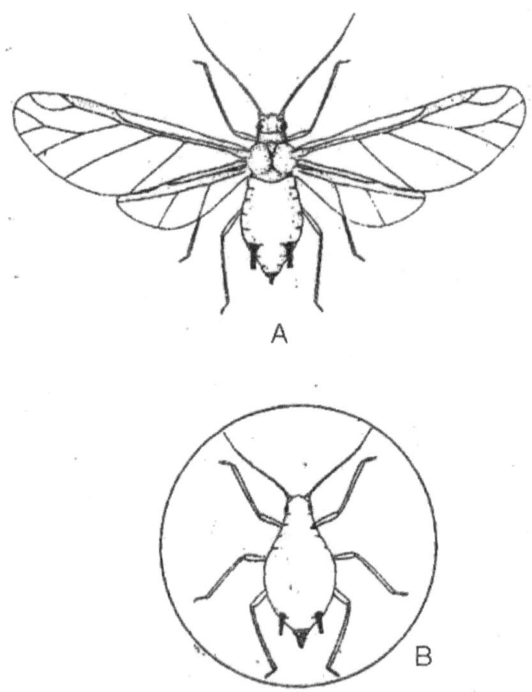

Fig. 138. Cotton aphid, *Aphis gossypii*
A - Adult. **B.** Nymph (From Hill, 1975)

Life history

Both the winged and wingless forms reproduce parthenogenetically and are viviparous. Each aphid may produce 8-22 offspring per day in its life of 2-3 weeks. A nymph moults four times before it becomes adult. One life cycle is completed in 7-9 days.

Control

The pest can be controlled by use of dimethoate @ 600 ml/ha or malathion @ 1000 ml/ha or monocrotophos @ 800 ml/ha.

10. COW BUG
Tricentrus bicolour Dist.
(HOMOPTERA : MEMBRACIDAE)

It is a membracid homopteran, commonly known as tree hopper. It is distributed throughout in India and is a minor pest of a variety of plnts like chilli, sugar-cane, redgram, alfalfa etc. The bug has a black body. Pronotum is subtriangular, overlapping the prothorax. There are two hollow hornlike superhumeral arms giving the appearance of a dagger when viewed laterally. The apex is projected posteriorly like a spine as is the case with other members of the family. Both the nymphs and adults suck the sap from the axils of branches. They excrete honey dew which invites ants. Damage to the stem is done by egg laying and not by sucking.

Life history

The adult bugs mate during warmer season. Eggs are deposited on the tender branches, petiole, or midrib of a leaf which are not directly exposed to the sun. The female cuts a longitudinal slit on the bark and insert eggs. Eggs may be laid singly or in groups. They hatch in about 8-12 days. The insect under goes five moults before it becomes adult. Newly hatched nymphs are soft bodied and shining. The colour of the nymphs in first reddish-brown which changes to brown with lateral dark bands. The nymphal period varies from 1-2 months.

Control

It is a minor pest and no economic damage is done to the host plants. The pest can be controlled by spraying with dimethoate @ 600 ml/ha or phosphamidon @ 300 ml/ha or malathion @ one litre/ha, particularly on the stem.

11. BEETLE GRUBS.
Anomala bengalensis. Blanch. *Holotrichia consanguinea* Blanch. and *Holotrichia reynaudi* Blanch., (= *H. insularis)*
(COLEOPTERA : MELOLONTHIDAE) (Figs. 139, 140)

The larvae of the above beetles are known as white grubs which have been reported to damage chilli plants. Out of the above,

Holotrichia reynaudi is common and very destructive pest of the chilli plants particularly in Rajasthan and other areas where the soil is sandy. The general features of all the white grubs are similar as already described for *Holotrichia consanguinea.*

Life Histroy

Eggs are laid in soil and the grubs remain confined to the soil and feed on roots, while the adult beetles hide in debris or in the soil during the day but come out at night and feed on leaves. The incubation period of *H. reynaudi* ranges from 8-12 days. The grubs remain active for 11-16 weeks and then overwinter in soil in the pupal stage. When the climatic conditions are favourable the beetles may emerge from the pupae in 2-3 weeks and later overwinter as adults. There is only one generation in a year.

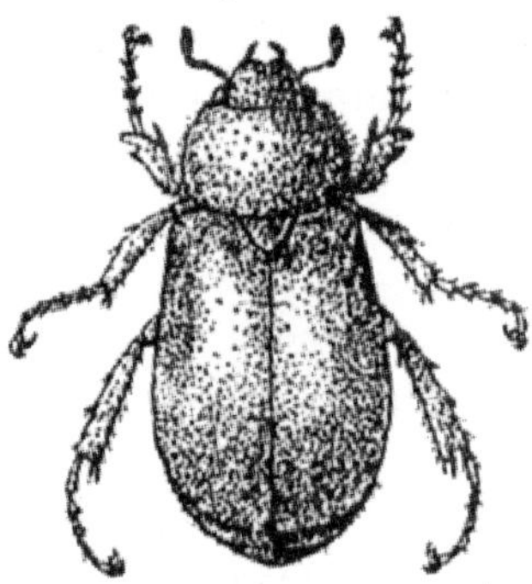

Fig. 139. *Holotrichia reynaudi* Adult
(From Srivastava and Butani, 2009)

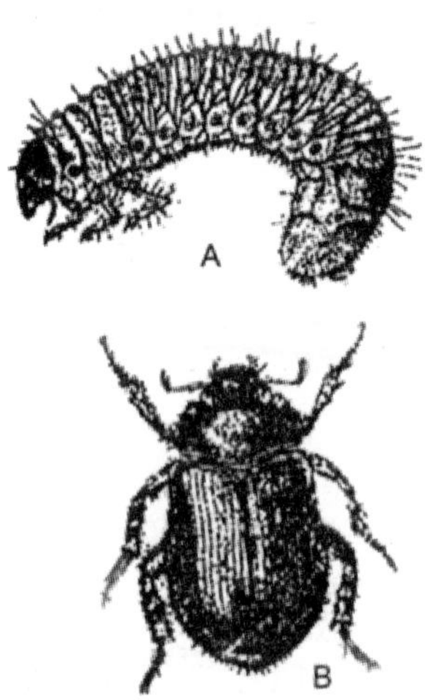

Fig. 140. *Anomala bengalensis* **A.** Grub, **B.** Adult
(From Srivastava and Butani, 2009)

Control

The pests can be controlled by 5% carbaryl or 10% phorate granules or qunolphos dust @ 20-22 kg/ha.

12. CHILLI FRUIT BORER
Longitarsus nigripennis Mots.
(CHRYSOMELIDAE : COLEOPTERA) (Fig. 141)

The chilli fruit borer is a serious pest of pepper, and has also been found damaging chillies.

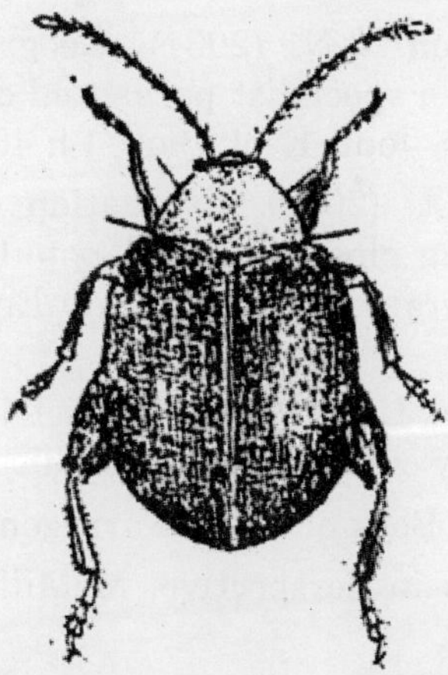

Fig. 141. Chilli fruit borer *Longitarsus nigripennis* Motsch. (From Srivastava and Butani, 2009)

Life history

Eggs are laid on chilli fruits and the grubs bore inside and feed within the fruits. The adult beetles scrap leaves as well as chilli fruits. The egg stage lasts 5-8 days, grubs 20-30 and pupal stages last 6-7 days after which the adult emerges. The total life cycle takes about 40-50 days. There are at least 4-overalpping generations in a year. The pest over winters in adult stage.

Control

Spraying 0.1% lindane or 0.2% carbaryl will control *L. nigripennis* as well as other fruit boring pest.

SELECTED BIBLIOGRAPHY

Althoff, D.M. and Thompson, J.N. (2001). Geographic Structure in the searching behaviour of a specialist parasitoid combining molecular and behavioural approaches. Jour. Evol. Biol., 14: 406-417.

Alvarez, J.M. and Hoy, M.A. (2002). Evaluation of ribosonal ITS2 DNA sequences in separating closely related populations of the parasitoid *Ageniaspis* (Hymenoptera: Encyrtidae). Ann Entomol. Soc. America. 95: 250-257.

Andrewartha, H.G. and Birch, L.C. (1954). The distribution and abundance of animals. The University of Chicago Press, Chicago.

Anonymous (2008). Amazing Book of Facts. Parragon Books, U.K.

Atkins, M.D. (1978). Insects in perspective. McMillan Publishing Inc. New York.

Atwal, A.S. (1976). Agricultural pests of India and South-East Asia, Kalyani Publisher, New Delhi.

Atwal, A.S. and Dhaliwal, G.S. (2012). Agricultural pests of South Asia and their management. Kalyani Publishers, Ludhiana, New Delhi etc.

Awasthi, V.B. (1969) Studies on the neurosecretory system of the active and hibernating adults of the house cricket, *Gryllodes sgillatus* (Walk) Anat Anz. 125 : 256-272.

Awasthi, V.B. (1976) Contribution to the neurohaemal organs in pterygote insects. Z. Mikrosk. Anat. Forsch, 90, 47-78.

Awasthi, V.B. (1979). Hormonal control of metabolism in insects. Indian J. Zoot. 20 : 1-38.

Awasthi, V.B. (1985). Influence of juvenile hormone analogue on the metamorphosis of *Philosamia ricini* Hutt. (Lepidoptera : Saturniidae). Entomon, 10 : 165-169.

Awasthi, V.B. (1985). Present status of insect Neuroendocrinology in India. Ind. Rev. Life. Sci. 5 : 59-91.

Awasthi, V.B. (2009). Introduction to general and applied entomology 3rd Edition, Scientific Publishers (India), Jodhpur, India.

Awasthi, V.B. (2013). Principles of Insect Behaviour. 2nd revised and enlarged edition. Scientific Publishers (India), Jodhpur, India.

Awasthi, V.B. (2005). Insect Neuroendocrines. Scientific Publishers (India), Jodhpur, India.

Awasthi, V.B. and Singh, U.V. (1983). The anatomy and histology of the ecdysial glands of *Amsacta collaris* Hamp. (Lepidoptera : Arctiidae) during postembryonic development. Zool. Jb. Anat. 109 : 467-480.

Ayyar, T.V.R. (1940). Hand book of economic entomology for South India, Govt. Press, Madras.

Barbosa, P. (1998). Conservation Biological Control. Academic Press, San Diego.

Beeson, C.F.C. (1941). The ecology and control of the forest insects of India and the neighbouring countries. Govt. India Press, Vasant Press, Dehra Dun.

Beroza, M. (1970). Chemicals controlling insect behaviour. Academic Press, New York.

Best, G.A. and Ruthven, A.D. (2005). Pesticides : Development, impact and controls. Woodland Publishing Limited, England.

Bhatnagar- Thomas, P.L. (1972). Laboratory evaluation of synthetic juvenile hormone analogue for the control of *Trogoderma granarium* Everts. Indian J. Entomol., 34 : 87-93.

Brain, M.V. (1983). Social insect ecology and behavioural biology. Chapman and Hall, London and New York.

Bray, D.P., Brandi, K.K., Brazil, R.P., Oliveira, A.G. and Itamiitan, J.G.E. (2009). Synthetic sex phromone attracts the leish maniasis vector. Lutzomyia longipaper (Diptera psychodidar) to trap in the field. Jour. Med. Environ. 46 : 428-434.

Briggs, G.G. (1994). Advances in the chemistry of insect control. III. Royal Society of Chemistry, Cambridge, U.K.

Brooks, G.T. and Roberts, T.R. (2005). Pesticide chemistry and bioscience. The food environment change. Woodland Publishing Limited, England.

Bryan, Beirne, P. (1967). Pest Management. Leonard Hill Books, London.

Burgess, N.R.H. (1990). Public health pests. Chapman and Hall, London.

Burt, A. (2003). Site specific selfish genes as tools for the control and genetic engineering of natural populations. *Proc. Roy. Soc. B.* 270 : 921-928.

Butler, C.D., Beckage, N.E. and Trumble, J.T. (2009). Effect of terrestrial pollutants on insect parasitoids. *Envir. Toxic. Chem.* 28: 1111-1119.

Chahal, G.S. and Gosal, S.S. (2002). Poincycles and procedures of plant breading. Biotechnological and conventional Approaches. Narosa Publishing House New Delhi.

Chand, P. (1995). Agricultural and forest pests and their management. Oxford and IBH Publishing Co. Pvt. Ltd. New Delhi.

Chapman, H.C. (1974). Biological control of Mosquito larvae. Ann. Rev. Entomol. 19 : 33-57.

Chapman, R.F. (1972). The Insects : Structure and Function, ELBS Edition, English University Press, London.

Chapman, R.F. and Boer, G. De. (1995). Regulatory mechanisms in insect Feeding. Chapman and Hall, London.

Clausen Curtis, P. (1962). Entomophagous Insects, Hafner Publishing Company, New York, U.S.A.

De Bach, P. (1964). Biological control of insect pests and weeds. Reinhold Publ. Corp. New York, U.S.A.

De Bach, P. (1974). Biological control by natural enemies. Cambridge University Press, Cambridge.

De Bach, P. and David, R. (1991). Biological control by natural enemies. Cambridge University Press, Cambridge.

Dent, D. (1991). Insect pest management. CAB international, Wallington, U.K.

Edwards, O.R. and Hoy, M.A. (1995). Monitoring Laboratory and field biotypes of the walnut parasite, *Trioxys palladus*, in a population cages using RAPD-PCR. Biocontrol Science and Technology, 5: 313-327.

Fabian, D.M., Alvarez, J.M. and Landis, D.A. (2004). Molecular Techniques and habitat manipulation approaches for parasitoid conservation in an croping systems. In : Ecological engineering for pest management (Gurr, G.M. Warattan, S.D., Atteri, M.A. eds.). CABI Publishing.

Fletcher, T.B. (1914). Some south Indian Insects, Govt. Press, Madras.

Gangrade, G.A. (1974). Insects of soybean. Technical bulletin number 24, J.N.K.V.V., Jabalpur, Madhya Pradesh.

Gatto, B., Zocca, A.; Battistu, A., Barrento, M.J., Braweo, M. and Paiva, M.R. (2009). Economic assessment of managing processionary moth in pine forest. A ease study in Portugal. J. Environ. Manag. 90 : 683-691.

Gillott, C. (1995). Entomology 2nd Edition, Plenum Press, New York.

Goodden, R. (1977). The wonderful world of Butterflies and Moths. Hamlyn.

Gullan, P.J. and Cranston, R.S. (1994). The insects: An outline of Entomology. Chapman and Hall, London.

Gullan, P.J. and Cranston, R.S. (2005). The Insects: An outline of Entomology, 3rd edn. Blackwell Publishing Ltd.

Gurr, M.G., Wratten, S.D. and Altieri, M.A. (2004). Ecological engineering for pest management: Advances in habitat manupulation for Arthropods. CABI Publishing.

Handler, A.M. and James, A.A. (2002). Insect transgenesis : Methods and applications, C.R.C. Press, U.S.A.

Harper, D.R. (2012). Viruses: Biology Applications Control. Garland Science, New York.

Highnam, K.C. and Hill, L. (1979). The comparative endocrinology of the Invertebrates. ELBS edition.

Hill, D.S. (1975). Agricultural insect pests of tropics and their control, Cambridge University Press, Cambridge.

Hill, D.S. (1994). Agricultural entomology, Timber Press, Portland, Oregon.

House, H. L. (1961). Insect nutrition. Ann. Rev. Entomol. 6, 13-26.

House, H. L. (1962). Insect nutrition. Ann. Rev. Biochem. 31, 563-672.

Howse, P. Stevens, I. and Jones, O. (1995). Insect pheromons and their use in pest management, Intercept Ltd. Andover Hants, U.K.

Hoy, M.A. (1994). Insect Molecular Genetics: An introduction to principles and applications. Academic Press, San Diego.

Huffaker, C.B. (1974). Biological control. Plenum Publishing Corporation, New York.

Imms, A.D. (1963). A general text book of entomology. Revised by O.W. Richards and R.G. Davies, Asia Publishing House, New Delhi.

Imms, A.D. (1964). A general text book of entomology. Revised by O.W. Richards and R.G. Davies. Methuen and Co. Ltd. London, E.P. Dutton and Co. Inc., New York.

Jacob, M. and Prabhu, V.K.K. (1979). Effects of two juvenile hormone analogues on embryonic neuroendocrine system of *Dysdercus cingulatus*. Heteroptera : Pyrrhocoridae) Entomon, 4 : 205-208.

Jacob, M. and Prabhu, V.V.K. (1979). Morphogenetic abnormalties caused by treatment of eggs of *Dysdercus cingulatus* with juvenile hormone analogues Curr. Sci., 48 : 505-507.

Jaw, J.H. and F.E. Regnier (1971). Pheromones. Ann. Rev. Biochem, 40 : 533-548.

Karlson, P. and Butendt, A. (1957). Pheromones (ectohormones) in insects. Ann. Rev. Entomol. 4: 39-58.

Kilgore, W.W. and Doutt, R. (1967). Pest control : Biological, Physical and Selected Chemical Methods. Academic Press, New York.

Koul, O. and Wahab, S. (2004). Neem : Today and in the new millennium. Kluwer Academic Publishers, Dordrect, Boston, London.

Landis, D.A. and Haas, M.J. (1992). Influence of landscape structure on abundance and within field distribution of *Ostrinia nubilais* Hubner (Lepidoptera: Pyradidae) Larval parasitoid in Michingen. Environmental Entomol. 21: 409-416.

Lefroy, H.M. (1909). Indian insect life. Thacker, Spink and Company, Calcutta.

Little, V.A. (1967). General and applied entomology. Oxford and IBH Publishing Company, Bombay and New Delhi.

Loxadale, H.D. and den Hollander, J. (1989). Electrophoretic studies on agricultural pests. Systematic Association Special, vol. 39. Clarendon Press, New York.

Mani, M.S. (1968). General Entomology. Oxford IBH Publishing Co., New Delhi.

Maramorosch, K. and Meinstosh A. H. (1994). Insect cell Biotechnology. Intercept Ltd. Andover, Hants, U.K.

Menken, S.B.J. and Ulenberg, S.A.C. (1987). Biochemical characters in Agricultural entomology. Agricul. Zool. Rev. 2: 305-360.

Menn, J.J. and Beroza, M. (1972). Insect Juvenile hormones : chemistry and action. Academic Press Inc. New York and London.

Metcalf, C.L. and Flint, W.P. (1962). Destructive and useful insects : Their habits and control. McGraw Hill Book Company, New York.

Metcalf, R.L. and Luckmann, W.H. (1994). Introduction to insect pest management, 3rd edition. John Wiley and Sons Inc., New York.

Nakanishi, K., Koreed, M., Sasaki, S., Chang, M.L. and Hsu, H.Y. (1966). The structure of ponasteron-A, an insect moulting hormone from the leaves of *Podocarpus nakaii*, Hay. Chem. Commun, 915-917.

Narang, S.K., Leopold, R.A., Krueger, C.M. and DeVault, J.D. (1994). Dichotomous RAPD-PCR key for identification of four species of parasitic hymonoptera. In: Applications of Genetics to Arthropods of Biological Control Significance. (S.K. Narang, A.C. Bartlett and R.M. Faust, eds.) pp. 53-67. CRC Press, Boca Raton, Florida.

Narwal, S.S. Tauro, P. and Bisla, S.S. (1997). Neem in sustanable agriculture, Scientific Publishers (India), Jodhpur, India.

Nayar, K.K., Ananthakrishnan, T.N. and David, B.V. (1976). General and applied entomology. Tata McGraw Hill, India.

Obrycki, J.J. and Tring, T. Jr. (1998). Predoceous Coccinellidae in biological control. Ann. Rev. Entomol. 43: 214-229.

Pal, M. (2002). Basics of Agriculture. Jain Brothers, New Delhi.

Patton, R.L. (1963). Introductory insect physiology. Topan and Company Limited, Tokyo.

Pfadt Robert, E. (1962). Fundamentals of applied Entomology. The McMillan and Company, New York.

Pimentel, D. (1981). Hand book of pest management in agriculture. Vol 1, 2 and 3, C.R.C. Press, U.S.A.

Pradhan, S. (1969). Insect pests of crops. National Book Trust of India, New Delhi.

Pradhan, S. (1983). Insect pests of crops revised by M.G. Jotwani, National Book Trust of India, New Delhi.

Price, P.W. (1975). Insect Ecology, John Wiley and Sons, London.

Prinsloo, G., Chen, Y., Giles, K.L. and Greenstone, M.H. (2002). Release and recovery in South Africa of the exotic aphid parasitoid *Aphelnus hordei* verified by the polymerase chain reaction. Bio Control 47: 127-136.

Pruthi, H.S. (1969). Text book on agricultural entomology. Indian Council of Agricultural Research, New Delhi.

Pruthi, H.S. and Singh, M. (1950). Pests of stored grains and control. Govt. of India Press.

Raabe, M. (1989). Recent developments in insect neurohormones. Plenum Press, New York.

Reay, R.C. (1969). Insects and insecticides. Oliver and Boyd, Edinburgha.

Rechigl, J.E. and Rechigl, N.A. (1998). Biological and Biotechnological control of insect pests, Lewis Publishers, U.S.A.

Reineke, A. Karlovsky, P. and Lebitz, C.P.W. (1999). Suppression of randomly primed polymerase chain reaction products (random amplified polymorphic DNA) in heterozygous diploid. Molecular Ecology. 8: 1449-1455.

Riddiford, L.M. (1970). Prevention of metamorphosis by exposure of insect eggs to juvenile hormone analogues. Science, 167 : 287-288.

Riddiford, L.M. and Williams, C.M. (1967). The effect of Juvenile hormone on the embryonic development of silkworm. Proc. Nat. Acad. Sci. (Amer.), 57 : 595-601.

Riddiford, L.M. and Williams, C.M. (1967). Volatile principle from oak leaves: Role in sex life of polyphemus moth. Science, New York 155 : 589-590.

Robbins, W.E., Kaplanis, J.N., Thompson, M.J., Shortino, T.J., Cohen, C.F. and Joyner, S.C. (1968). Ecdysones and analogues: effects on development and reproduction of insects. Science, 161 : 1158-1159.

Robinson, W.H. (1996). Urban Entomology, Chapman and Hall, London.

Roeder, K.D. (1953). Insect physiology. John Wiley and Sons, Inc. New York.

Roonwal, M.L. (1979). Termite life and termite control in tropical south Asia. Scientific Publishers (India), Jodhpur, India.

Rosen, D. and DeBach, P. (1973). Systematics, morphology and biological control. Entomophaga. 18: 215-222.

Ruberson, J.R., Nemoto, H. and Hirose, Y. (1998). Pesticides and conservation of natural enemies in pest management. In Conservation Biological Control (P. Barbosa, ed.) pp. 207-220. Academic Press, San Diago.

Russel, A.S.H. (2007). Whitaker's World of Facts. *Penguin Books.*

Russell, T.L. and Kay, B.H. (2008). Biologically used insecticides for the control of immature Australian mosquitoes: a review. Austr. J. Entomol. 47 : 232-242.

Ryan, L. (1995). Post harvest tobacco infestation control. Chapman and Hall, London.

Samways, M.J. (1995). Insect conservation biology. Chapman and Hall, London.

Schneidermann, H.A. (1972). Insect hormones and insect control. Insect Juvenile Hormones : Chemistry and Action. Academic Press, New York.

Schowalter, T.D. (2011). Insect Ecology : An ecosystem Approach. 3rd edition, Elsevier-Academic Press

Shatters, S. R.G.Jr. (2000). Environmental impact of biotechnology. In : Biological and Biotechnological control of insect pests. (Recheigl, J.E. and Recheigl, N.A. eds.) Lewis Publishers, New York.

Sidall, J.B. (1970). Chemical aspects of hormonal interactions. Chemical Ecology. pp. 281-306. Academic Press Inc., New York.

Singh, G. and Prakash, S. (2009). Effect of *Bacillus sphaericus* against larvae of malaria and filarial vectors : an analysis of early resistance detection, Parasitol. Ros. 104 : 763-766.

Singh, H., Kumar, V., Kumar, R. and Rohilla, H.R. (1997). Stored grain insect pest management. In : Neem in sustanable agriculture. Scientific Publishers (India), Jodhpur.

Singh, N.P. and Singh, R.A. (2002). Scientific Crop Production with reference to north eastern hill region. Kalyani Publishers, Ludhiana.

Slama, K. and Williams, C.M. (1965). Juvenile hormone activity for the bug, *Pyrrhocoris apterus*. Proc. Nat. Acad. Sci (Amer.) 54 : 411-414.

Slama, K. and Williams, C.M. (1966). The juvenile hormone, V. The Sensitivity of the bug *Pyrrhocoris apterus* to a hormonally active factor in American Paper Pulp. Biol. Bull. Woods Hole, 130 : 235-246.

Slama, K. and Williams, C.M. (1967). Paper Factor, as an inhibitor of the embryonic development in the European bug, *Pyrrhocoris apterus*. Nature, 210: 329-330.

Slama, K., Ramanuk, M. and Sorm, F. (1974). Insect hormones and bioanalogues. Springer Verlag, Wien New York.

Snodgrass, R.E. (1935). Principles of insect morphology. McGraw Hill, Company, New York.

Srivastava, K.P. and Butani, D.K. (2009). Pest Management in vegetables. Vol. 1. 2nd edn., Studium Press (India) Pvt. Ltd., New Delhi.

Staal, G.B. (1971). Practical aspects of insect control by juvenile hormone. Bull. W.H. Org. 44 : 391-394.

Symondson, W.O.C. and Liddell, J.E. (1996). The ecology of agricultural pests, Biochemical approach. Chapman and Hall, London.

Truman, J.W. and Riddiford, L.M. (1974). Hormonal mechanisms underlying insect behaviour. Adv. Ins. Physiol. 10 : 297-352.

Uvarov, B. (1966). Grasshoppers and Locusts : A hand book of general Acridology. Vol. 1. Cambridge University Press.

Vaughn, T.Y.T. and Antolin, M.F. (1998). Population genetics of an opportunistic parasitoid in an agricultural landscape. Heredity, 80: 152-162.

Walia, S. and Parmar, B.S. (1995). Pesticides, crop protection and environment. Oxford and IBH Publishing Co. Pvt. Ltd., New Delhi.

Walter, G.H. (2003). Insect Pest Management and Ecological Research. Cambridge University Press, Cambridge.

Welling, W. (1977). Dynamic aspects of insects, insecticide interactions. Ann. Rev. Entomol. 22 : 153-178.

West, T.F. and Hard, I.E. (1961). Chemical control of insects. Chapman and Hall Ltd., London.

Whitten, M.J. and Foster, G.G. (1975). Genetical methods of pest control. Ann. Rev. Entomol. 20 : 461-476.

Wigglesworth, V.B. (1965). The principles of insect physiology. ELBS edition. Methuen and Company, London.

Wigglesworth, V.B. (1969). Insect physiology. Chapman and Hall, London.

Williams, C.M. (1967). Third generation pesticides, Scientific American, 217 : 13-17.

Wood, J.G.R. (1994). Insects of the world. Vol. 1 and 2 Atlantic Publishers and distributors, New Delhi.

Yamanaka, T. and Liebhold, A.M. (2009). Spatially implict approach to understand the manipulation of mating success for insect invasion management. Population Ecol. 51 : 427-444.

Zhu, Y.C. and Greenstone, M.H. (1999). Polymerase chain reaction techniques for distinguishing three species and two strains of *Aphelinus* (Hymenoptera: Aphinidae) from *Diuraphis noxia* and *Schizaphis graminum* (Homoptera : Aphididae). Ann. Entomol. Soc. Amer. 92: 71-79.

APPENDIX - 1

THE COMMON NAMES AND TRADE NAMES OF SOME INSECTICIDES

Common name	Trade name
Azinphosmethyl	Gusathion or Guthion
Carbaryl	Sevin, Carvin, Carbaryl
Carbofenothion	Trithion, Garrathion
Diazinon	Basudin
Dichlorvos (DDVP)	Vapona, Nuvan, Vaporan
Dicofol	Kelthane, Dicofol, Delcofol
Dimethoate	Rogor, Dimetox, Dimethos
Demeton methyl (Methyl demeton)	Metaxsystox
Endosulfan	Thiodan, Endosol, Agrosulfan
Fenitrothion	Folithion, Sumithion, Accothion
Formothion	Anthio
Malathion	Malatox, Cythion
Menazon	Sayfos
Methyl parathion	Folidol, Dalf, Paradon
Monocrotophos	Nuvacron, Corophos, Croton, Monocos
Mevinphos	Phosdrin
Phorate	Thimet, Foratox
Phosphamidon	Dimecron, Daron, Cildon
Thiometon	Ekatin
Trichlorphon/Trichlorfon	Dipterex, Tugon
Quinalphos	Desalux, HLX, Kinalux

APPENDIX - II

Here a brief account of only 28 principal insect orders has been given which have been arranged according to convenience in the tables alongwith their agricultural importance.

Characteristics of Insect orders

S.N.	Order	Appearance	Examples	Metamorphosis	Wings	Mouth parts	Antennae	Tarsal segments	Habitat	Agricultural importance injurious[1]
1	Protura		Telsontails	None	None	Sucking	None	1	Terrestrial and semi aquatic	—
2	Thysanura	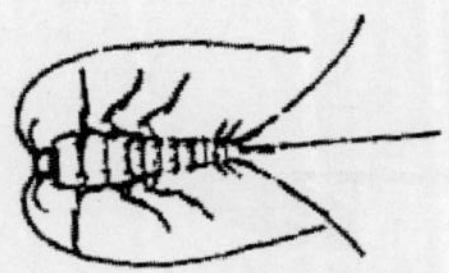	Silverfish and allies	None	None	Chewing	Long, many segmented	2-3	Terrestrial and semi aquatic	—
3	Collembola		Springtails	None	None	Chewing	Short to long	fused to tibia or 1	Terrestrial and semi aquatic	+
4	Diplura	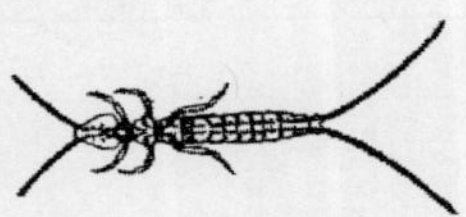	Japygids	None	None	Chewing	Short to long	1	Terrestrial	—

1 Dash indicates minor agricultural importance; pluses indicate extent of major agricultural importance.
(Modified after Pfadt, 1962)

Characteristics of Insect orders

S.N.	Order	Appearance	Examples	Metamorphosis	Wings	Mouth parts	Antennae	Tarsal segments	Habitat	Agricultural importance injurious[1]
5	Ephemeroptera		Mayflies	Simple	2 pr. membranous	Chewing	Bristle like, short	3-5	Aquatic as nymphs	—
6	Odonata		Dragonflies, damsel flies	Simple	2 pr. mem branous	Chewing	Bristle like, short	3	Aquatic as nymphs	—
7	Plecoptera		Stoneflies	Simple	2 pr. mem branous	Chewing	Thread-like, long	3	Aquatic as nymphs	+
8	Orthoptera	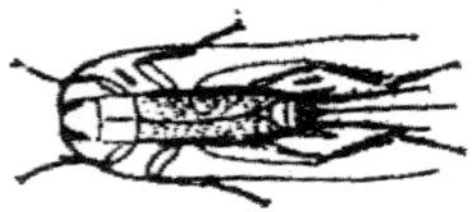	Grasshoppers, crickets, etc.	Simple	Tegmina & membranous	Chewing	Thread-like, short to long	0-5	Terrestrial	++++
9	Dictyoptera or Blattaria	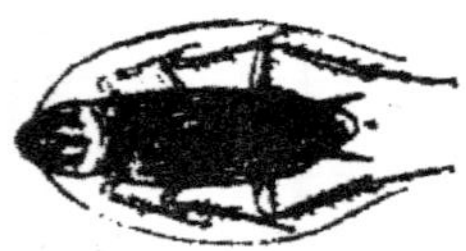	Cockroachs, mantids etc.	Simple	2 pr. membranous or none	Chewing	Thread-like, long	0-5	Terrestrial	++
10	Phasmida		Leaf insects, stick insect, etc.	Simple	2 pr. or none	Chewing	Long or short	0-5	Terrestrial	+

Characteristics of Insect orders

S.N.	Order	Appearance	Examples	Metamorphosis	Wings	Mouth parts	Antennae	Tarsal segments	Habitat	Agricultural importance injurious[1]
11	Dermaptera	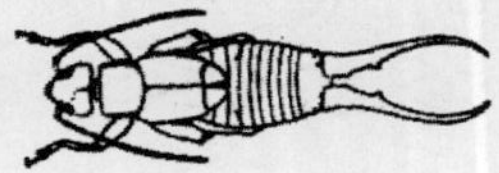	Earwigs	Simple	Short tegmina & membranous	Chewing	Thread-like, long	3	Terrestrial	++
12	Embioptera	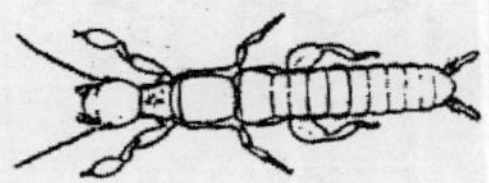	Embiids	Simple	2 pr. membranous or none	Chewing	Thread-like	3	Terrestrial and semi aquatic	—
13	Isoptera		Termites	Simple	2 pr. membranous or none	Chewing	Beadlike, short to long	4-5	Terrestrial	++
14	Psocoptera		Booklice and psocids	Simple	2 pr. membranous or none	Chewing	Thread-like, generally long	2-3	Terrestrial	—
15	Mallophaga		Biting lice, bird lice	Simple	None	Chewing	Thread-like, or clublike and short	1-2	Terrestrial, cetoparasitic	++

Characteristics of Insect orders

S.N.	Order	Appearance	Examples	Metamorphosis	Wings	Mouth parts	Antennae	Tarsal segments	Habitat	Agricultural importance injurious[1]
16	Anoplura	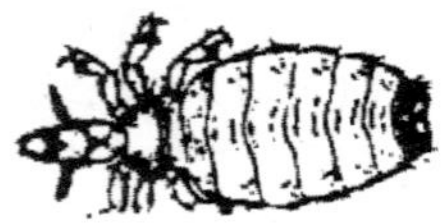	Sucking lice	Simple	None	Piercing sucking	Bristle like, short	1	Terrestrial, ecto-parasitic	+++
17	Thysanoptera	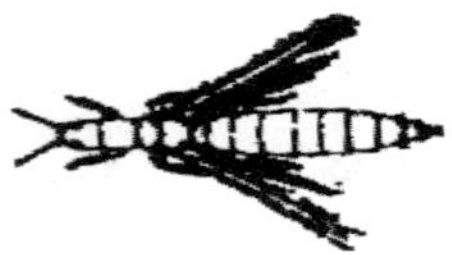	Thrips	Simple	2 pr. slender fringed with hair	Rasping-sucking	Short	1-2	Terrestrial	+++
18	Heteroptera		Bugs	Simple	Hemelytra & membranous or none	Piercing-sucking	Thread-like, short to long	2-3	Terrestrial and aquatic	++++
19	Homoptera		Aphids, scales, leaf hoppers	Simple	2 pr. membranous or none	Piercing - sucking	Bristle like, or thread like	1-3	Terrestrial	+++++
20	Neuroptera		Lacewings and allies	Complex	2 pr. membranous	Chewing	Usually thread like	5	Terrestrial and aquatic	—

Characteristics of Insect orders

S.N.	Order	Appearance	Examples	Metamorphosis	Wings	Mouth parts	Antennae	Tarsal segments	Habitat	Agricultural importance injurious[1]
21	Coleoptera		Beetles, weevils	Complex	Elytra and membranous	Chewing	Variable, short to long	1-5	Terrestrial and aquatic	+++++
22	Strepsiptera		Twisted-winged insects, stylopids	Complex	Front vestigeal, hind membranous, none in female	Chewing	Short, none in female	2-5	Terrestrial most parasitic in insects	—
23	Mecoptera		Scorpion-flies	Complex	2 pr. membranous or none	Chewing	Thread-like, long	5	Terrestrial	—
24	Trichoptera		Caddis-flies	Complex	2 pr. membranous with hairs or scales	Chewing	Thread-like, long	5	Aquatic as larvae and pupae	—
25	Lepidoptera		Butter-flies, moths, skippers	Complex	2 pr. membranous with scales	Siphoning, chewing in larvae	Variable, short to long	5	Terrestrial	+++++

Characteristics of Insect orders

S.N.	Order	Appearance	Examples	Metamorphosis	Wings	Mouth parts	Antennae	Tarsal segments	Habitat	Agricultural importance injurious[1]
26	Hymenoptera		Ants, bees, wasps, etc.	Complex	2 pr. membranous or none	Chewing or chewing-lapping	Usually thread-like, elbowed	5	Terrestrial	++
27	Diptera		Flies	Complex	membranous and halteres or none	Piercing sucking or spon-ging; chewing in larvae	Thread-like or aristate, usually short	5	Terrestrial many aquatic larvae and pupae	+++++
28	Siphonaptera		Fleas	Complex	None	Piercing-sucking, chewing in larvae	Knobbed and short	5	Terrestrial ecto-parasitic as adults	++

INDEX